博物馆典藏版

THE BIRDS OF AMERICA

美洲鸟类

[美] 约翰·詹姆斯·奥杜邦 (John James Audubon) / 绘

张劲硕 吴海峰 / 校译

U0233785

人民邮电出版社

北京

图书在版编目（CIP）数据

美洲鸟类 : 博物馆典藏版 / (美) 约翰·詹姆斯·奥杜邦 (John James Audubon) 绘 ; 张劲硕, 吴海峰校译. -- 北京 : 人民邮电出版社, 2018.2
ISBN 978-7-115-46998-4

Ⅰ. ①美… Ⅱ. ①约… ②张… ③吴… Ⅲ. ①鸟类—美洲—图谱 Ⅳ. ①Q959.708-64

中国版本图书馆CIP数据核字(2017)第279779号

版权声明

This edition of The Birds of America was first published in England in 2012 by The Natural History Museum, London.
Copyright © Natural History Museum, 2012.
This edition is published by Posts and Telecommunications Press by arrangement with the Natural History Museum, London.

◆ 绘 [美] 约翰•詹姆斯•奥杜邦(John James Audubon)
校 译 张劲硕 吴海峰
责任编辑 韦 毅
责任印制 彭志环
◆ 人民邮电出版社出版发行 北京市丰台区成寿寺路 11 号
邮编 100164 电子邮件 315@ptpress.com.cn
网址 http://www.ptpress.com.cn
北京捷迅佳彩印刷有限公司印刷
◆ 开本：880×1230 1/16
印张：29 2018 年 2 月第 1 版
字数：637 千字 2018 年 2 月北京第 1 次印刷
著作权合同登记号 图字：01-2014-4926 号

定价：258.00 元
读者服务热线：(010)81055410 印装质量热线：(010)81055316
反盗版热线：(010)81055315

前 言

1804年，19岁的约翰·詹姆斯·奥杜邦（John James Audubon）站在美国宾夕法尼亚州的一个洞穴中，观察着一对灰胸长尾霸鹟，他顿悟到："我专心致志地观望着它们天真自由的样子，这时，有个想法就像一道灵光在我的脑海中一闪而过，毕竟，没有什么事情能让我为了表现大自然而愿意全身心地投入，唯有以大自然自己的方式去临摹它，展现它生机勃勃和感人至深的一面！我曾经连续数月观察鸟类，无论它们是在降落或是在飞翔，并绘制了上百幅鸟类素描图。"

由此可见，那个时期的奥杜邦对鸟类绘画已从着迷变成痴迷。他步入北美洲的荒野，在森林中、河流旁到处寻找鸟类，采集标本、研究、整姿，并画下他见到的每一种鸟类。多年之后，他的绘画作品结集出版，永远地改变了鸟类学和博物艺术的相关领域。

对于现在的观鸟者来说，在奥杜邦所处的那个年代，他所遇到的各种困难是难以想象的。除了摆在面前的贫困生活——饥寒交迫、蚊虫叮咬、环境脏乱，他还没有照相机和望远镜，更没有其他鸟友或是鸟类野外观察手册。他带着一把猎枪和绘画工具，凭借学到的与鸟类有关的一些零散知识，通过自己的观察，而且仅仅是简单的观察，去尝试发现鸟类形态之间细微的异同。

奥杜邦几乎将自己的精力全部投入到鸟类身上，这使得他多次商场失意，并使家庭生活陷入困境，但他（和他那同样不屈不挠的妻子露西）总能找到办法渡过难关。我们通过他的信件和日记可以感受到他的激情以及熠熠生辉的敬业精神。1810年，奥杜邦帮助他的生意伙伴经营一家干货店，就在那时他记述了在肯塔基州路易斯维尔的时光，他写道："我打猎，我绘画，我只是为了观察大自然，我的想法与他人不同，除了这些我什么都不在乎。"

在奥杜邦之前的鸟类绘画基本上表现的都是呆板的姿态，而奥杜邦在19岁时给自己定下的目标是：让活生生的鸟类跃然纸上，重塑鸟类优雅、美丽的姿态。但他仍用了很多年才打破传统，并走出了一条自己的路。

在第141页的下方有两只鸟——左侧是苍鹰，右侧是库氏鹰。这幅作品创作于1809年左右，就在他顿悟之后第5年，这幅作品显示出他对于传统呆板的绘画和干瘪的姿态只是作了些许的修饰而已。20年后，即1829年，在他的作品终稿刚刚出版之后，他又在图版上方补画了更有活力且更生动的苍鹰的亚成体。

将故事情节和鸟类行为引入鸟类绘画的创意打破了科学与艺术的界限，甚至超越了二者的界限。1902 年，美国博物学家约翰·巴勒斯（John Burroughs）曾写下这样的评价："(他的作品）太过于注意感情的流露，甚至有时太过夸张，表演性和戏剧性的味道太浓。"他指责奥杜邦绘画中的戏剧效果会有失准确性，而且奥杜邦很显然对自己讲述的故事十分沾沾自喜，并且加以了润色。

当奥杜邦创作小嘲鸫保卫鸟巢的作品时（第 21 页），他选择了一个十分夸张的捕食者——木纹响尾蛇，它的上半身离开树枝，探到巢中。巴勒斯随即指出了他的错误，木纹响尾蛇不会爬树。但与此同时，奥杜邦一丝不苟地把小嘲鸫画得非常准确，英勇无畏的小嘲鸫保卫自己爱巢的瞬间被完美地捕捉到了，表现了这种鸟精力充沛、活跃好动的特性。

奥杜邦的作品中经常采取的一种艺术手法是将多个事件融入一幅作品中。在他的那幅描绘黄胸大鹏莺的作品中（第 137 页），一只雌鸟卧于巢中，但却有 3 只雄鸟或飞或栖于它的上方，其实这个场景对于这种独来独往的鸟类来说似乎不太可能出现。如果我们推断奥杜邦其实是想让我们看到一种"延时画面"，即实际上只有一只雄鸟在鸟巢上方炫耀，然后为雌鸟饲喂食物的话，那么这幅作品讲述的故事就变得顺理成章了。这幅作品向我们讲述的是作为个体的鸟的生活细节，但从鸟类学层面上讲也算得上是准确的。

在人们所知的所有奥杜邦的绘画作品中，最令人激动的一幅表现的是一只赤肩鵟凶猛地捕食一群惊慌失措、四散逃窜的山齿鹑（第 76 页）。这只赤肩鵟的姿态看起来虽然稍微有点儿别扭，但这幅作品的每个细节都是各有道理的，好像赤肩鵟被记录的瞬间，正是左脚在右脚之前，而它的右翅是在左翅之前。这是一幅时间上不连贯的作品，它反映的不是一瞬间的画面，而是把多个时间发生的事件融入一幅作品中；这幅作品有效地记录了动作爆发的瞬间，并且能让人感受到同时而来的声响。

值得注意的是，对于当代欣赏者来说，这只赤肩鵟看起来别扭的原因是我们的直觉已经被一生中所看到的照片影响了。奥杜邦和他同时代的人只能用自己的双眼来观察世界，像一只鹰袭击一群鹑类那样转瞬即逝的瞬间本就是模糊不清的画面。奥杜邦只能想象细节，这些想象对于这幅作品或是其他作品都是正确的。可以推断，他给鸟类整姿要尽可能自然，要与他多年积累下来的观鸟经验所留下的印象相匹配。作为一个艺术家，他也必须根据当下的流行趋势，下意识地选择那些优雅、生动并且最具吸引力的姿态。

但对于所有类似的图片中要传达的剧情与暴力，每只鸟都被精美的手工绘画渲染出了细节。正当鸟儿们在为最重要的生死时刻而奋斗，画面被定格在一瞬间。奥杜邦鼓励我们去欣赏它们的形态和图案之美。

奥杜邦非常在意欣赏者对他作品的评价，这使他需要投入更多的金钱继续他的工作，乔纳森·罗森（Jonathan Rosen）发现"奥杜邦总是在同一个时间干很多事情，为了他的订购者紧赶慢赶着他的巨大项目，为自己做广告，制作一个野性美洲的浪漫版。”他特意留起长头发，涂抹着熊油，让头发下垂且油光滑亮。他拜访英国皇家学会的时候还要穿着鹿皮大衣。英国人着迷于这位“美国林人”，对他在荒野的生活故事、不可思议的鸟类绘画非常有兴趣。他总是有意识或无意识地去描绘鸟类最引人入胜和最原初的瞬间，他要反映出一个自然仍然占绝对优势的世界，他就是要用自己的绘画强调大自然的无穷魅力。

这就好像奥杜邦在他的作品中一直要表达的那样：它们不是生活在英国乡间的小鸟，而是分布于美洲那广袤而未知的原野中的鸟类，我走进荒野，在我的画纸上驯服它们，捕捉住它们的特点，这样你就可以欣赏它们的美了。毫无疑问，奥杜邦的作品在英国比在美国更有市场，随着美洲的原野今非昔比，他的作品在大众中变得越来越流行。

奥杜邦花了12年多的时间，全身心地投入到他的创作中，他的作品成为了不朽之作。他画鸟，写书，监制图版的雕刻和包装，把作品卖给美国和欧洲的出版商，有时甚至将完成的作品直接打包寄给订购者。在这些工作中，他的妻子、两个儿子、很多朋友和助手给了他莫大的帮助，从写作、编辑文本到打理生意，甚至包括在图版上绘制植物和背景。

最终，奥杜邦共为1050多只鸟类绘制了同比例大小的图画，出版了435幅图版。除此之外，他还写了5卷《鸟类学记述》，其中有很多他自己对鸟类习性的观察记录。他从佛罗里达到拉布拉多，从得克萨斯到达科他，一路旅行，并且8次横渡大西洋。

重新审视奥杜邦的成就甚为紧迫。1839年，在《美洲鸟类》完成之后，一家波士顿的报纸对该书写下了这样的评论：“它完成了一次无与伦比的大胆、几近鲁莽的突破，它承载着一种毅力和不懈的热情，而且它体现了忠诚度、行业规范和快速有效性，对美国的商界和科学界来说，它将是一座不朽的丰碑。”这本书对一个一心一意、下定决心的男人来说是一座重要的里程碑，他的传奇故事让绘画这门艺术更具吸引力。

时至今日，这些绘画作品仍有巨大的魅力，不难想象，在19世纪早期的英国，它们肯定也具有很大的吸引力，并创造了在任何其他国家都不会出现的激动人心的历史。每一幅绘画作品都是美洲原野早期的一幅照片。当然，它们有一些夸张和过度渲染，但每幅作品都讲述了一个故事，不只是鸟类的生活史，还有奥杜邦探索荒野时的兴奋和好奇之情。每一点都能将我们带回到小木屋或被蜡烛照亮的客厅，抑或是奥杜邦发现了想象和再创作鸟类故事乐趣的原野。谁不愿意用心倾听一段美妙的故事呢？

大卫·阿伦·西布利

（David Allen Sibley）

美国著名鸟类学家

译者序

奥杜邦的每一幅作品，无论是在艺术创作层面的绘画手法、表现形式，还是在博物发现层面的对鸟类结构或细节的观察、对行为动作的刻画，在经历了两百年的沉淀之后，都早已凝固成不可替代的经典。

如今，博物学几乎成为显学，出版界亦是推手之一，遂可见有为数不少的奥杜邦的图书得以问世，并且版本众多，赞美之词不绝于耳。

然而，当我们接手人民邮电出版社的翻译任务之初，便希望呈现一个更科学、准确的译本，并对奥杜邦的作品做一客观的评价——尽管我们未必有资格。

画法与画作

受制于当时的技术条件，在对鸟类进行观察和描绘时，奥杜邦没有能够定格瞬间的照相机，有的只是一沓画纸、一把画笔和一杆猎枪。虽然奥杜邦可以把鸟儿拿在手中摆弄姿态，或制成标本仔细地观察它的结构细节，但他仅凭一双肉眼，无法准确地观察到某些鸟儿活着时候的姿态，尤其是转瞬即逝的运动状态，更无法准确地将这些姿态像拍照片那样呈现在纸上。

因此，奥杜邦作品中某些鸟儿的运动姿态可能稍显“扭曲”和“生硬”，这与今天我们看到的“鸟类野外观察手册”风格的科学画有较大不同，这貌似成为了奥杜邦作品的一种风格、一种时代造就的特殊的“科学画”。

尽管奥杜邦对鸟类的大多数细节有准确的把握，但我们在仔细研究他的作品之时，仍能发现些许瑕疵，尤其他对颜色的处理，有的存在一定问题。这也许是因为奥杜邦观察的那只鸟儿当时的状态本身就与我们在图鉴或照片中看到的“典型”存在差别，这只鸟可能刚刚经过了繁殖期，它的羽毛或许已经有些磨损，或是变得暗淡；也可能是奥杜邦在绘画之时，对于某些颜色的把握存在偏差；更重要的原因是奥杜邦的很多画作可能根本就是照着死去的标本绘就的，姿态和颜色都发生了改变；抑或是后人在翻版上色的过程中对颜色的处理没有严格遵照原版，因而产生了颜色过深、过浅或对比度过大、过小等情况。

实际上，市面上流行的不同版本间某些颜色的偏差不仅是因为纸张与印刷的差别。我们看到的绝大多数奥杜邦的作品都不是他的原作手稿，也不是对原作的影印，而是经翻刻铜版后印制而成的。因此，我们看到的作品中的颜色，大都为后人所上，而可能与奥杜邦当时所见及所绘之颜色有所差别，这一点看似无足轻重，但实际上却对某些鸟种的鉴别构成了足够大的困难。

除了奥杜邦之后画匠的上色和印刷工艺可能导致了我们现在看到的作品的颜色与初稿有别之外，奥杜邦本人似乎也喜欢为一些鸟儿"画蛇添足"，但这对于纯粹的物种鉴别来讲却没有起到"画龙点睛"的作用。例如，东王霸鹟（*Tyrannus tyrannus*）（第 79 页）头的上部本应为灰黑色，但奥杜邦却将它的"模特"头顶正中央点缀上了红色；而叉尾王霸鹟（*Tyrannus savana*）（第 168 页）头上部本应为蓝黑色，奥杜邦却将其顶冠画成了黄色；与之相反的是，美洲树雀鹀（*Spizelloides arborea*）（第 188 页）胸前本应有一个深棕色斑点，但在奥杜邦的画笔下这只鹀胸前的斑点却神秘地消失了！虽然这些鸟儿的"特点"凭空产生或消失了，但凭借奥杜邦对其绘画对象身体其他细节的准确把握，我们还是能够确定它们的具体种类。

分类与鉴定

两百年过去了，人们对一些鸟种的认识发生了变化，尤其是那些基于形态解剖学和分子生物学的对鸟类分类地位的调整：人们对一些鸟所属的分类单元——属（genus）进行了厘定，而这与奥杜邦当时的认知存在差别；另有一些鸟种，由于进一步的研究而被划分为两个或多个鸟种，而奥杜邦进行绘画创作时，他显然不可能区别所有的近缘种、同一种的不同亚种或至今我们还没有确定的隐蔽种的关键特征。奥杜邦观点上的"局限"受制于时代，或许两百年后的人们也会否定我们现在对这个世界的某些认知。

在确定奥杜邦绘画作品中鸟类的具体种类时，我们不但参考了美国奥杜邦协会网站、美国国家地理学会出版的《北美鸟类野外手册（第 6 版）》（*National Geographic Field Guide to the Birds of North America, Sixth Edition*）等权威机构发布的鉴定资料，还参考了国内之前出版发行的多个版本的《美洲鸟类》（*the Birds of America*）。

在翻译鸟种名称时，我们主要采用中国科学院院士、北京师范大学教授郑光美先生主编的《世界鸟类分类与分布名录》中的信息对奥杜邦作品中鸟种的中文名、学名及英文名进行标注。我们希望读者通过标注中的信息，不但能以现代的分类视角来认识当年绘画作品中的鸟种，还可以利用这些信息，在互联网或其他书籍中对感兴趣的鸟种进行深入的检索、查询与了解。

但郑先生的著作出版于 2002 年，在此后的 15 年间，科学家对鸟类分类的研究工作从未止步，奥杜邦作品中的一些鸟类的分类地位也发生了变化。为了使读者能够更方便地在国外网站或较新的文献中查找到目标鸟种的信息，对于这些分类地位发生变化的鸟种，我们还采用了国际鸟类学委员会（International Ornithology Committee，IOC）整理的 2015 年版的《世界鸟类名录》（*IOC World Bird List, V5.4*）中的学名和较为统一的英文正规名。最终，对于在郑先生主编的名录中和 IOC 名录中学名不同的鸟种，我们标注出 IOC 中更新的学名，并在紧随其后的括号中标注出郑先生书中原来的学名或属名。

应当指出的是，郑先生著作中的学名在该书出版发行至今的这段时间内一直为国内绝大多数鸟类学相关资料所采用，但实际上仍然有比郑先生观点更加“古老”的观点；IOC 虽然体现了最新的分类系统，但随着研究的进一步深入，仍然有更新的分类学观点不断涌现，这些观点或许将逐渐为学界所采纳。但我们为读者提供的郑先生版和 IOC 2015 年版的鸟种学名，或许在相当长一段时期内，仍将是主流。

标注与翻译

纵观之前国内的多个版本的《美洲鸟类》，对于一些鸟种的中文名、学名和英文名标注存在不一致的情况。如果是因为译者采用了不同的分类观点，或是对奥杜邦所绘的某些鉴别特征不明显的鸟种认知的差异而造成了这些不同，尚情有可原。但不可接受的是，对于某些鸟种，一些版本采用了直译奥杜邦标注的英文名称的办法。例如，奥杜邦将一种啄木鸟命名为“Maria’s Woodpecker”，某些版本的译者直接将其译为“玛利亚的啄木鸟”，而且类似的情况还不止一处，这就失去了译者标注的作用。我们甚至认为，“译者”的工作并不是对奥杜邦原作中的文字信息进行简单的“翻译”，而是应该像上文中提到的那样逐一查阅、鉴别。

除此之外，还有一个在之前几个版本中都比较常见的问题。国际动物命名法委员会（International Commission on Zoological Nomenclature，ICZN）于 2000 年颁布的最新版本的《国际动物命名法规》（*International Code of Zoological Nomenclature*, Appendix B General Recommendations, 6）明确指出，属或种组的分类单元的学名须以与正文不同的字体印刷。这些名称通常印成斜体，但是不可用在较高分类单元的名称上。种组名称必须以小写字母起首，当引证时须以属名在先（或缩写），所有高于种的分类单元均须以一大写字母起首。

而学界一般是这样处理这个问题的：当物种的英文名为正体时，若由一个词组成，应仅首字母大写，其余字母均为小写；若英文名由两个或多个独立的词组成，每个词均以大写字母起首，其余字母均为小写；对于带连字符的词，连字符后面的一个词全部字母均为小写。而对于学名，由于是拉丁文，根据上述法规，则应以与正体英文名相区别的斜体标注，属名（第一个词）仅首字母大写，而种本名或种加词（第二个词）全部小写。

但实际上，在之前每个版本的《美洲鸟类》中，我们都能发现不符合上述规范的问题。这虽然貌似无伤大雅，但却违背了法规的原则，同时也给一些细心的读者带来了困惑。这将不利于科学的传播，也有悖于法规的初衷。而这种问题是完全可以避免的。因此，在您手捧的这本由人民邮电出版社出版的《美洲鸟类（博物馆典藏版）》中，我们尽力在中文的翻译部分注意了规范性，其余部分仍采用原作的处理方式。

对原作的处理

市面上流行的几个版本的《美洲鸟类》或许都曾受到读者的追捧，它们不同的纸张与印刷效果为读者带来不同的视觉体验。其中某些版本，或许是为了追求所谓的“整体效果”而抹去了奥杜邦本人当年在绘制作品时所做的文字标注，还有些版本或许是为了追求所谓的“艺术效果”而对原画进行了剪裁、旋转。我们无意评判孰优孰劣，况且这些处理或许能够被部分读者所接受，但它们毕竟已经不是原始的、完整的奥杜邦的作品，而这些被“切割”了的作品也已经失去了一部分原作品传达给我们的信息。因此，我们没有对原作进行任何涂抹、删除或裁剪等处理，而是完整地保留了原作的全部内容。如此一来，我们不仅可以看到在其他版本中或许看不到的奥杜邦的文字批注，甚至还可以通过当年的文字，对那时人们对自然的认知以及动物分类学观点窥之一二。

最后，希望读者通过本书可以更好地了解奥杜邦及其作品，也请广大读者对书中存在的疏漏或错误给予批评指正。

张劲硕　吴海峰

2016 年 10 月

目 录

火 鸡（雄性）/ 1
黄嘴美洲鹃 / 2
蓝翅黄森莺 / 3
紫朱雀 / 4
加拿大威森莺 / 5
火 鸡（雌性和雏鸟）/6
拟八哥 / 7
白喉带鹀 / 8
黑枕威森莺（雌性或亚成体）/9
黄腹鹨 / 10
白头海雕（雄性，亚成体）/ 11
橙腹拟鹂 / 12
暗眼灯草鹀 / 13
草原林莺 / 14
北森莺 / 15
游 隼 / 16
哀 鸽 / 17
比氏苇鹪鹩 / 18
白眉灶莺 / 19
蓝翅虫森莺 / 20
小嘲鸫 / 21
紫崖燕 / 22
黄喉地莺 / 23
黄喉地莺 / 24
歌带鹀 / 25
卡罗莱纳（长尾）鹦鹉/26
红头啄木鸟 / 27
蓝头莺雀 / 28
棕胁唧鹀 / 29
松 莺（雄性）/ 30
白头海雕（雄性）/ 31
黑嘴美洲鹃 / 32
美洲金翅雀 / 33
食虫莺 / 34
黄林莺 / 35
库氏鹰 / 36
北扑翅鴷 / 37
黄腹地莺 / 38
美洲凤头山雀 / 39
橙尾鸲莺 / 40
披肩榛鸡 / 41
圃拟鹂 / 42
雪松太平鸟 / 43
玫红丽唐纳雀 / 44
桤木纹霸鹟 / 45
横斑林鸮（成年雄性）/46
红喉北蜂鸟 / 47
白喉林莺 / 48
白喉林莺（雄性）/ 49
纹胸林莺（雄性）/ 50
红尾鵟 / 51
卡氏夜鹰 / 52
丽彩鹀 / 53
刺歌雀 / 54
红冠戴菊（雄性）/ 55
赤肩鵟 / 56
呆头伯劳 / 57
隐夜鸫 / 58
栗胁林莺 / 59
美洲雕鸮 / 61
旅 鸽（已灭绝）/ 62
白眼莺雀（雄性）/ 63
沼泽带鹀（雄性）/ 64
黄林莺（雄性）/ 65
象牙喙啄木鸟 / 66
红翅黑鹂 / 67
美洲燕 / 68
栗胸林莺 / 69
亨氏草鹀 / 70
赤肩鵟（成年雄性）/ 71
燕尾鸢（雄性，雌雄同色）/ 72
棕林鸫 / 73
靛彩鹀 / 74
灰背隼（雄性）/ 75
山齿鹑 / 76
白腹鱼狗 / 77
卡罗苇鹪鹩 / 78
东王霸鹟 / 79
黄腹鹨 / 80
鹗（雄性）/ 81
三声夜鹰 / 82
莺鹪鹩 / 83
灰蓝蚋莺 / 84
黄喉林莺（雄性）/ 85
红尾鵟 / 86
丛 鸦 / 87
栗胸林莺 / 88
黄喉虫森莺 / 89
黑白森莺 / 90
巨翅鵟 / 91
灰背隼 / 92
海滨沙鹀 / 93
栗肩雀鹀（雄性）/ 94

黄林莺 / 95
白喉鹊鸦 / 96
东美角鸮 / 97
长嘴沼泽鷦鷯 / 98
褐头牛鹂 / 99
双色树燕 / 100
渡　鸦（雄性）/ 101
冠蓝鸦 / 102
加拿大威森莺 / 103
棕顶雀鹀（雄性）/ 104
红胸䴓 / 105
黑头美洲鹫 / 106
灰噪鸦 / 107
狐色雀鹀 / 108
稀树草鹀 / 109
黑枕威森莺 / 110
北美黑啄木鸟 / 111
绒啄木鸟 / 112
东蓝鸲 / 113
白冠带鹀 / 114
东绿霸鹟（雄性）/ 115
褐弯嘴嘲鸫 / 116
密西西比灰鸢 / 117
歌莺雀 / 118
黄喉莺雀（雄性）/ 119
灰胸长尾霸鹟 / 120
雪　鸮 / 121
斑翅蓝彩鹀 / 122
纹胸林莺 / 123
黑头威森莺 / 124
褐头䴓 / 125
白头海雕（亚成体）/ 126
玫胸白斑翅雀 / 127
灰嘲鸫 / 128
大冠蝇霸鹟（雄性）/ 129
黄胸草鹀（雄性）/ 130
旅　鸫 / 131
黑背啄木鸟 / 132
白颊林莺 / 133
橙胸林莺 / 134
橙胸林莺（雄性）/ 135
东草地鹨 / 136
黄胸大鹏莺 / 137
灰喉地莺 / 138
田雀鹀（雄性）/ 139
松　莺 / 140
苍　鹰 / 141
库氏鹰（成体）/ 141
美洲隼 / 142
橙顶灶莺 / 143
绿纹霸鹟 / 144
棕榈林莺 / 145
鱼　鸦 / 146
美洲夜鹰 / 147
黑喉蓝林莺 / 148
尖尾沙鹀 / 149
红眼莺雀（雄性）/ 150
红头美洲鹫 / 151
白胸䴓 / 152
黄腰林莺 / 153
灰冠虫森莺（雄性）/ 154
黑喉蓝林莺（雄性）/ 155
短嘴鸦（雄性）/ 156
锈色黑鹂 / 157
烟囱雨燕 / 158
主红雀 / 159
卡罗山雀 / 160
凤头巨隼 / 161
鸣哀鸽 / 162
棕榈林莺 / 163
棕夜鸫（雄性）/ 164
巴氏猛雀鹀（雄性）/ 165
毛脚鵟（雄性）/ 166
绿顶鹑鸠 / 167
叉尾王霸鹟（雄性）/ 168
红树美洲鹃（雄性）/ 169
灰王霸鹟 / 170
仓　鸮 / 171
蓝头鹑鸠 / 172
家　燕 / 173
绿胁绿霸鹟 / 174
短嘴沼泽鷦鷯 / 175
枞树镰翅鸡 / 176
白顶鸽 / 177
橙冠虫森莺 / 178
莺鷦鷯（雄性）/ 179
松金翅雀 / 180
金　雕（成年雌性）/ 181
地　鸠 / 182
金冠戴菊 / 183
黑喉芒果蜂鸟 / 184
黑胸虫森莺 / 185
草原松鸡 / 186
宽尾拟八哥 / 187
美洲树雀鹀 / 188
雪　鹀 / 189
黄腹吸汁啄木鸟 / 190
柳雷鸟 / 191
灰伯劳 / 192
林氏带鹀 / 193
北山雀 / 194
红冠戴菊 / 195
矛　隼 / 196
红交嘴雀 / 197
白眉食虫莺 / 198
棕榈鬼鸮 / 199
角百灵 / 200
加拿大黑雁 / 201
红喉潜鸟 / 202
王秧鸡 / 203
长嘴秧鸡 / 204
弗吉尼亚秧鸡 / 205
林鸳鸯 / 206
褐鲣鸟 / 207
极北杓鹬 / 208
厚嘴鸻 / 209
姬苇鳽 / 210
大蓝鹭（雄性）/ 211
环嘴鸥 / 212
北极海鹦 / 213
刀嘴海雀 / 214
红颈瓣蹼鹬 / 215
黑头鹮鹳 / 216
三色鹭（成年雄性）/ 217
崖海鸦 / 218
白翅斑海鸽 / 219
笛　鸻 / 220
绿头鸭 / 221
美洲白鹮 / 222
美洲蛎鹬 / 223
三趾鸥 / 224
双领鸻 / 225
美洲鹤（成年雄性）/ 226
针尾鸭 / 227
绿翅鸭 / 228
小潜鸭 / 229

三趾滨鹬 / 230
长嘴杓鹬 / 231
棕胁秋沙鸭 / 232
黑脸田鸡 / 233
环颈潜鸭 / 234
乌燕鸥 / 235
夜　鹭 / 236
中杓鹬 / 237
云斑塍鹬 / 238
美洲骨顶 / 239
粉红燕鸥 / 240
大黑背鸥 / 241
雪　鹭（春季的成年雄性）/ 242
扇尾沙锥 / 243
黑水鸡（成年雄性）/ 244
厚嘴崖海鸦 / 245
欧绒鸭 / 246
斑脸海番鸭 / 247
斑嘴巨鹱鹛 / 248
簇羽海鹦 / 249
北极燕鸥 / 250
褐鹈鹕（成年雄性）/ 251
角鸬鹚（成年雄性）/ 252
中贼鸥 / 253
细嘴瓣蹼鹬 / 254
灰瓣蹼鹬 / 255
棕颈鹭 / 256
角鸬鹚（春季的成年雄性）/ 257
棕塍鹬 / 258
角鸊鷉 / 259
白腰叉尾海燕 / 260
沙丘鹤（亚成体）/ 261
白尾鹲 / 262
弯嘴滨鹬 / 263
暴风鹱（夏季的成年雄性）/ 264
黄胸鹬 / 265
普通鸬鹚 / 266
长尾贼鸥 / 267
小丘鹬 / 268
青脚鹬 / 269
黄蹼洋海燕 / 270
华丽军舰鸟（成年雄性）/ 271
短尾贼鸥 / 272
橙嘴凤头燕鸥（春季的成年雄性）/ 273
斑翅鹬 / 274
白顶玄燕鸥（成年雄性）/ 275
王绒鸭 / 276
加拿大黑雁 / 277
白腰滨鹬 / 278
白嘴端凤头燕鸥 / 279
黑浮鸥 / 280
大蓝鹭（白色型，春季的成年雄性）/ 281
冰岛鸥 / 282
大　鹱（雄性）/ 283
紫滨鹬 / 284
厚嘴鸻（左 3）/ 284
叉尾鸥 / 285
白额雁 / 286
白　鸥 / 287
小黄脚鹬（夏季的雄性）/ 288
褐腰草鹬 / 289
黑腹滨鹬 / 290
银　鸥 / 291
凤头鸊鷉 / 292
角海鹦 / 293
斑胸滨鹬 / 294
大西洋鹱（雄性）/ 295
白颊黑雁 / 296
丑　鸭 / 297
赤颈鸊鷉 / 298
奥氏鹱（春季的雄性）/ 299
美洲金鸻 / 300
帆背潜鸭 / 301
北美黑鸭 / 302
高原鹬 / 303
翻石鹬 / 304
紫青水鸡（春季的成年雄性）/ 305
普通潜鸟 / 306
小蓝鹭 / 307
大黄脚鹬 / 308
普通燕鸥（春季的雄性）/ 309
斑腹矶鹬 / 310
美洲鹈鹕（成年雄性）/ 311
长尾鸭 / 312
蓝翅鸭 / 313
笑　鸥 / 314
红腹滨鹬 / 315
美洲蛇鹈 / 316
斑头海番鸭 / 317
褐胸反嘴鹬 / 318
小白额燕鸥 / 319
美洲小滨鹬 / 320
粉红琵鹭（成年雄性）/ 321
美洲潜鸭 / 322
黑剪嘴鸥（雄性）/ 323
博氏鸥 / 324
白枕鹊鸭 / 325
北鲣鸟 / 326
琵嘴鸭 / 327
黑颈长脚鹬（雄性）/ 328
北美花田鸡（春季的成年雄性）/ 329
半蹼鸻 / 330
普通秋沙鸭 / 331
拉布拉多鸭（已灭绝）/ 332
美洲绿鹭 / 333
灰　鸻 / 334
短嘴半蹼鹬 / 335
黄冠夜鹭 / 336
美洲麻鳽 / 337
布氏鸭（绿头鸭和赤膀鸭的杂交种，12 月的雄性亚成体）/ 338
侏海雀 / 339
暴风海燕 / 340
大海雀（已灭绝）/ 341
鹊　鸭 / 342
棕硬尾鸭 / 343
高跷鹬 / 344
绿眉鸭 / 345
黑喉潜鸟 / 346
斑头秋沙鸭 / 347

赤膀鸭 / 348
黑田鸡 / 349
岩　鸧（成年雌性）/ 350
乌林鸮（成年雌性）/ 351
黑翅鸢 / 352
栗背山雀 / 353
黑顶山雀 / 353
短嘴长尾山雀 / 353
黄腹丽唐纳雀 / 354
猩红丽唐纳雀 / 354
灰头地莺 / 355
白尾鹞 / 356
黑嘴喜鹊 / 357
松　雀 / 358
棕腹长尾霸鹟 / 359
剪尾王霸鹟 / 359
西王霸鹟 / 359
鸊　鷉 / 360
岩异鹩（雌性）/ 360
蓝镰翅鸡 / 361
暗冠蓝鸦 / 362
黄嘴喜鹊 / 362
西丛鸦 / 362
北美星鸦 / 362
太平鸟 / 363
白翅交嘴雀 / 364
铁爪鹀 / 365
矛　隼（雌性）/ 366
斑尾鸽 / 367
岩雷鸟 / 368
高山弯嘴嘲鸫（雄性）/ 369
杂色鸫（雄性和雌性）/ 369
美洲河乌 / 370
艾草松鸡 / 371
斯氏鵟（雌性）/ 372
黄昏锡嘴雀 / 373
黑头白斑翅雀 / 373
纹腹鹰 / 374
白腰朱顶雀 / 375
黑嘴天鹅（亚成体）/ 376
秧　鹤 / 377
猛　鸮 / 378
棕煌蜂鸟 / 379
鬼　鸮 / 380
雪　雁 / 381
尖尾松鸡 / 382
长耳鸮（雄性）/ 383
美洲雀 / 384
崖沙燕 / 385
紫绿树燕 / 385
大白鹭（春季的雄性）/ 386
彩　鹮（成年雄性）/ 387
三色黑鹂（成年雄性）/ 388
黄头黑鹂 / 388
布氏拟鹂（成年雄性）/ 388
红顶啄木鸟 / 389
鹨雀鹀（雄性）/ 390
白斑黑鹀 / 390
歌带鹀（雄性）/ 390
黑　雁 / 391
栗翅鹰（成年雌性）/ 392
黄眉林莺（雄性）/ 393
山蓝鸲 / 393
西蓝鸲 / 393
栗领铁爪鹀（春季的雄性）/ 394
黑头白斑翅雀（老年雄性）/ 394
金冠带鹀（成年雄性）/ 394
棕胁唧鹀 / 394
黄腰林莺 / 395
黄脸林莺 / 395
黑喉灰林莺 / 395
北极鸥 / 396
美洲红鹮 / 397
白腹蓝彩鹀（春季的雄性）/ 398
褐雀鹀（雄性）/ 398
暗眼灯草鹀 / 398
黑喉绿林莺 / 399
橙胸林莺（雌性）/ 399
灰头地莺 / 399
暗背金翅雀（雄性）/ 400
白腰朱顶雀（雄性）/ 400
黄腹丽唐纳雀（雌性）/ 400
唐氏雀（雄性）/ 400
黄腹铁爪鹀（雄性）/ 400
红胸秋沙鸭 / 401
扁嘴海雀 / 402
小海雀 / 402
凤头海雀 / 402
角嘴海雀 / 402
巴氏鹊鸭（夏季的雄性）/ 403
黑颈䴙䴘 / 404
半蹼滨鹬 / 405
黑嘴天鹅（成体）/ 406
乌信天翁 / 407
黑海番鸭 / 408
弗氏燕鸥 / 409
白顶燕鸥 / 409
鸥嘴噪鸥（夏季的雄性）/ 410
小天鹅（亚成体）/ 411
海鸬鹚（冬季的雌性）/ 412
加州鸬鹚（雄性）/ 412
珠颈斑鹑 / 413
金翅虫森莺 / 414
栗颊林莺 / 414
美洲旋木雀 / 415
褐头鳾 / 415
长嘴啄木鸟 / 416
红腹啄木鸟 / 416
北扑翅䴕 / 416
刘氏啄木鸟 / 416
红胸吸汁啄木鸟 / 416
长嘴啄木鸟 / 417
三趾啄木鸟（美洲）/ 417
柳雷鸟（左，春季的雄性）/ 418
白尾雷鸟（右，冬季羽征）/ 418
隐夜鸫（雄性）/ 419
坦氏孤鸫（雌性）/ 419
灰噪鸦（雄性，亚成体）/ 419
红翅黑鹂 / 420
褐鹈鹕（第一年冬季的

亚成体）/ 421
毛脚鵟 / 422
冠齿鹑（假想雄性亚成体）/ 423
山翎鹑 / 423
白腹蓝彩鹀（雌性）/ 424
家朱雀（雄性）/ 424
褐头牛鹂（雄性）/ 424
粉红腹岭雀（雄性，亚成体）/ 424
黄昏锡嘴雀 / 424
狐色雀鹀（雌性）/ 424
安氏蜂鸟 / 425
加州神鹫（老年雄性）/ 426
北美蛎鹬 / 427
短嘴鹬（雌性）/ 428
小绒鸭 / 429
斑海雀 / 430
大红鹳（老年雄性）/ 431
穴小鸮（雄性）/ 432
纵纹腹小鸮（雌性）/ 432
北美鸺鹠（雄性）/ 432
短耳鸮（雄性）/ 432
橙腹拟鹂 / 433
黄脸金翅雀 / 433
杂色鸫（雌性）/ 433
黄眉灶莺（雄性）/ 433
小纹霸鹟 / 434
小头莺（雄性）/ 434
蓝山莺（雄性）/ 434
黄喉莺雀（雄性）/ 434
绿胁绿霸鹟（雄性）/ 434
黑长尾霸鹟（雄性）/ 434
美洲河乌 / 435

图版信息 / 436
索引 / 446

Great American Cock Male

Drawn by J. J. Audubon F.R.S.E. F.L.S. M.W.S.

Vulgo (*Wild Turkey*) MELEAGRIS GALLOPAVO

Engraved by W.H.Lizars Edinr.
Retouched by R. Havell Junr. London 1829

火 鸡（雄性）

学 名：*Meleagris gallopavo*

英文名：Wild Turkey

No.I. PLATE II.

Yellow-billed Cuckoo

COCCYZUS CAROLINENSIS

Plant Popaw Porceliatriloba

黄嘴美洲鹃

学　名：*Coccyzus americanus*

英文名：Yellow-billed Cuckoo

Drawn by J. J. Audubon F.R.S.E. F.L.S. M.W.S.

Engraved by W.H.Lizars Edinr.
Retouched by R. Havell Junr. London 1829

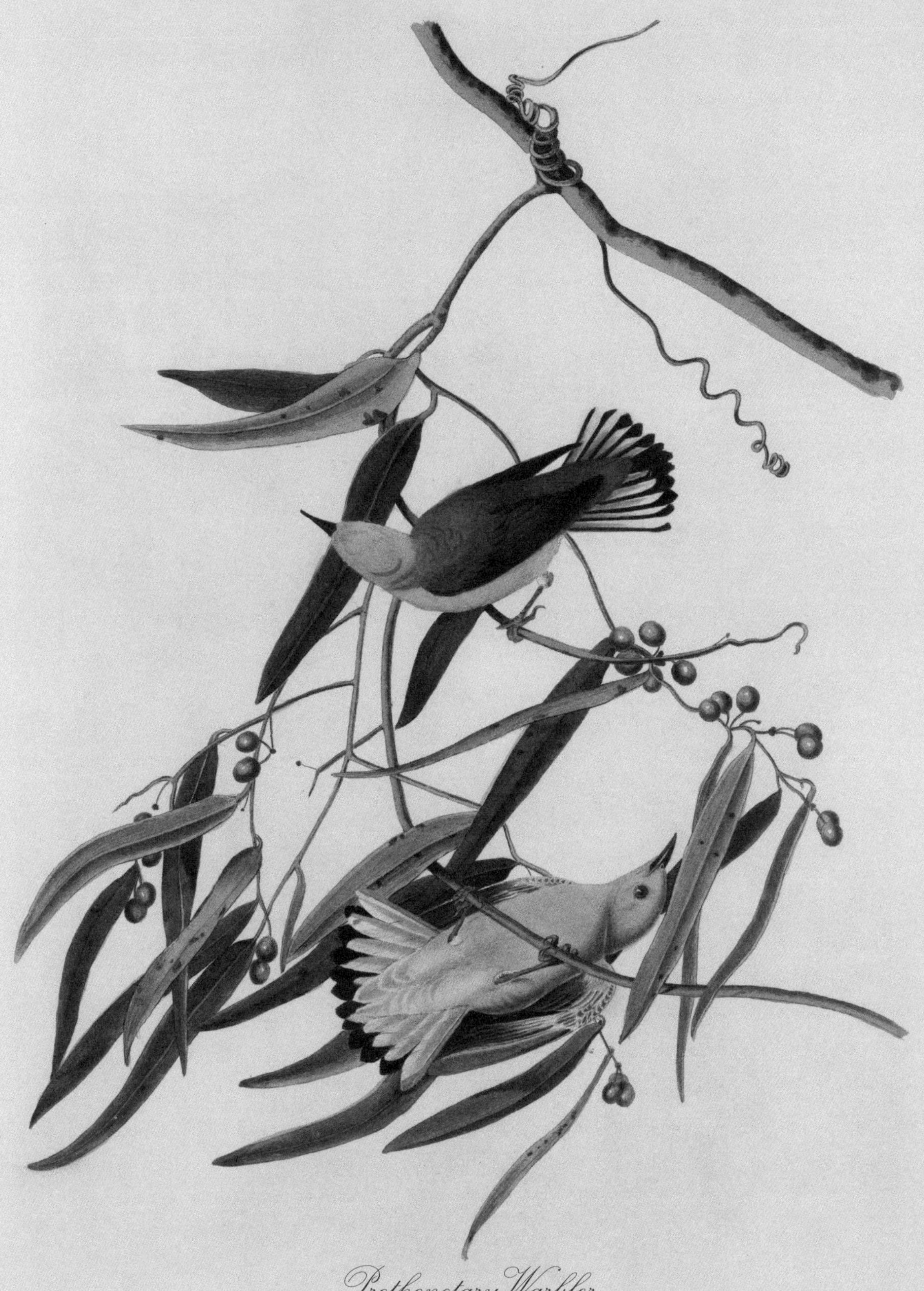

Prothonotary Warbler

DACNIS PROTONOTARIUS

Plant Vulgo Cane Vine

Drawn by J. J. Audubon M.W.S.

Engraved by W. H. Lizars Edinr.

蓝翅黄森莺

学　名：*Protonotaria citrea*

英文名：Prothonotary Warbler

Purple Finch

FRINGILLA PURPUREA

Plant Pinus pendula

Vulgo Black Larch

Drawn by J. J. Audubon M.W.S.

Engraved by W. H. Lizars Edinr.

紫朱雀

学　名：*Haemorhous purpureus*

（原为 *Carpodacus* 属）

英文名：Purple Finch

Bonaparte Fly Catcher

MUSCICAPA BONAPARTII

Drawn by J. J. Audubon M.W.S.

Plant seed pud Magnolia grandiflora

Engraved by W. H. Lizars Edinr.

加拿大威森莺

学　名：*Cardellina canadensis*

（原为 *Wilsonia* 属）

英文名：Canada Warbler

Nº. II. PLATE VI.

Great American Hen & Young

Drawn from Nature by John J. Audubon F.R.S.E. M.W.S. *Vulgo female Wild Turkey* MELEAGRIS GALLOPAVO Coloured by R. Havell Senr. Engraved by W.H.Lizars Edinr.

火　鸡（雌性和雏鸟）

学　名：*Meleagris gallopavo*

英文名：Wild Turkey

Purple Grackle

QUISCALUS VERSICOLOR

1. Male 2. Female

Plant Vulgo, Indian Corn

Drawn from Nature by John J. Audubon F.R.S.E. F.L.S. M.W.S.

Printed & coloured by R. Havell Senr.

Engraved by W.H.Lizars Edinr.
Retouched by R. Havell Junr. London 1829

拟八哥

学　名：*Quiscalus quiscula*

英文名：Common Grakle

1. 雄性　2. 雌性

White Throated Sparrow

FRINGILLA PENSYLVANICA

1. Male 2. Female

Plant Coruns Florida – Vulgo Dog Wood

Drawn from Nature by John J. Audubon F.R.S.E. M.W.S.　Printed & Coloured by R. Havell Senr.　Engraved by W.H.Lizars Edinr.

白喉带鹀

学　名：*Zonotrichia albicollis*

英文名：White-throated Sparrow

1. 雄性　2. 雌性

Selby's Fly Catcher

MUSCICAPA SELBII

Drawn from Nature by John J. Audubon F.R.S.E. M.W.S. *Plant Vulgo, Pheasants Eye* Printed & Coloured by R. Havell Senr. Engraved by W.H.Lizars Edinr.

黑枕威森莺（雌性或亚成体）

学　名：*Setophaga citrina*

（原为 *Wilsonia* 属）

英文名：Hooded Warbler

No 2.

PLATE X.

Brown Lark

ANTHUS AQUATICUS

1. Male 2. Female

Drawn from Nature by John J. Audubon F.R.S.E. M.W.S.

Printed and Coloured by R. Havell Senr.

Engraved by W.H.Lizars Edinr.

黄腹鹨

学　名：*Anthus rubescens*

英文名：Buff-bellied Pipit

1. 雄性　2. 雌性

The Bird of Washington or Great American Sea Eagle

Drawn from Nature by John J. Audubon F.R.S.E. M.W.S.

FALCO WASHINGTONIENSIS *Male*

Engraved by R. Havell Junr.
Printed and coloured by R. Havell Senr.

白头海雕（雄性，亚成体）

学　名：*Haliaeetus leucocephalus*

英文名：Bald Eagle

Baltimore Oriole *1 & 2. Males 3. Female and Nest*

ICTERUS BALTIMORE

Plant Vulgo, Yellow Poplar

Liriodendron Tulipifera

Drawn from Nature by John J. Audubon F.R.S.E. M.W.S. Printed & Coloured by R. Havell Senr. Engraved by R. Havell Junr.

橙腹拟鹂

学　名：*Icterus galbula*

英文名：Baltimore Oriole

1,2. 雄性　3. 雌性和巢

Snow Bird 1. Male 2. Female

FRINGILLA NIVALIS

Drawn from Nature by John J. Audubon F.R.S.E. M.W.S.　*Plant Vulgo. Great Swamp Ash*　Printed & Coloured by R. Havell Senr.　Engraved by R. Havell Junr.

暗眼灯草鹀

学　名：*Junco hyemalis*

英文名：Dark-eyed Junco

1. 雄性　2. 雌性

Prairie Warbler 1. Male 2. Female

SYLVIA DISCOLOR

Drawn from Nature by John J. Audubon F.R.S. M.W.S. *Plant Vulgo. Buffalœ Grafs* Printed & Coloured by R. Havell Senr. Engraved by R. Havell Junr.

草原林莺

学　名：*Setophage discolor*

（原为 *Dendroica* 属）

英文名：Prairie Warbler

1. 雄性　2. 雌性

Blue Yellow back Warbler (1. Male 2. F.)

SYLVIA AMERICANA

Drawn by J. J. Audubon F.R.S.E. M.W.S.

Plant, Vulgo. Louisiana Flag

Printed & Coloured by R. Havell Senr.

Engraved by R. Havell Junr.

北森莺

学　名：*Steophaga americana*

（原为 *Parula* 属）

英文名：Northern Parula

1. 雄性　2. 雌性

No.4. PLATE 16.

Great Footed Hawk Male 1. F. 2.

Drawn from Nature & Published by John J. Audubon F.R.S.E. M.W.S.

FALCO PEREGRINUS

Engraved, Printed & Coloured by R. Havell & Son, London

游　隼

学　名：*Falco peregrinus*

英文名：Peregrine Falcon

1. 雄性　2. 雌性

Carolina Pigeon or Turtle Dove Male 1. F. 2.

COLUMBA CAROLINENSIS

Drawn from Nature and Published by John J. Audubon F.R.S.E. M.W.S. *Plant Steuartia Malacodrendron* Engraved, Printed & Coloured by R. Havell & Son, London

哀 鸽

学 名：*Zenaida macroura*

英文名：Mourning Dove

1. 雄性 2. 雌性

Bewick's Long Tailed Wren

TROGLODYTES BEWICKII

Drawn from Nature by John J. Audubon F.R.S.E. M.W.S. *Plant Vulgo. Iron Wood* Printed & Coloured by R. Havell Senr. Engraved by R. Havell Junr.

比氏苇鹪鹩

学　名：*Thryomanes bewickii*

英文名：Bewick's Wren

Louisiana Water Thrush

TURDUS AQUATICUS

Drawn from Nature by John J. Audubon F.R.S.E. M.W.S.   *Plant Vulgo. Indian Turnip*   Printed & Coloured by R. Havell Senr.   Engraved by R. Havell Junr.

白眉灶莺

学　名：*Parkesia motacilla*

（原为 *Seiurus* 属）

英文名：Louisiana Waterthrush

Blue Winged Yellow Warbler Male 1. F. 2.

DACNIS SOLITARIA

Drawn from Nature by John J. Audubon F.R.S.E. M.W.S. *Plant Vulgo. Wild Althea* Printed & Coloured by R. Havell Senr. Engraved by R. Havell Junr.

蓝翅虫森莺

学　名：*Vermivora cyanoptera*

（原为 *Pinus* 属）

英文名：Blue-winged Warbler

1. 雄性　2. 雌性

The Mocking Bird 1. Male 2. F.

TURDUS POLYGLOTTUS

Play Vulgo. Yellow Jessamin

Rattlesnake

CROTALUS HORRIDUS

Drawn from Nature & Published by John J. Audubon F.R.S.E. M.W.S.

Engraved, Printed & Coloured by R. Havell & Son, London

小嘲鸫

学　名：*Mimus polyglottos*

英文名：Northern Mockingbird

1. 雄性　2. 雌性

木纹响尾蛇

学　名：*Crotalus horridus*

英文名：Timber Rattlesnake

Purple Martin 1. Male 2. F.

HIRUNDO PURPUREA

Drawn from Nature and Published by John J. Audubon F.R.S.E. M.W.S.

Nest, a Gourd

Engraved, Printed & Coloured by R. Havell & Son, London

紫崖燕

学　名：*Progne subis*

英文名：Purple Martin

1. 雄性　2. 雌性

Maryland Yellow Throat Male 1. F. 2.

SYLVIA TRICHAS

Plant Vulgo—1. Wild Olive—2. Bitter Wood.

Drawn from Nature and Published by John J. Audubon F.R.S.E. M.W.S.　　Engraved, Printed & Coloured by R. Havell & Son, London

黄喉地莺

学　名：*Geothlypis trichas*

英文名：Common Yellowthroat

1．雄性　2．雌性

Roscoe's Yellow Throat

SYLVIA ROSCO

Drawn from Nature and Published by John J. Audubon F.R.S.E. M.W.S.

Plant Vulgo. Swamp Oak

Engraved, Printed & Coloured by R. Havell and Son

黄喉地莺

学　名：*Geothlypis trichas*

英文名：Common Yellowthroat

Song Sparrow Male 1. F. 2.

FRINGILLA MELODIA

Drawn from Nature and Published by John J. Audubon F.R.S.E. M.W.S.

Plant Vulgo. Wortle Berry

Engraved, Printed & Coloured by R. Havell & Son, London

歌带鹀

学　名：*Melospiza melodia*

英文名：Song Sparrow

1. 雄性　2. 雌性

Carolina Parrot Males 1. F. 2. Young 3.

PSITACUS CAROLINENSIS

Drawn from Nature & Published by John J. Audubon F.R.S.E. M.W.S.　*Plant Vulgo. Cuckle Burr.*　Engraved, Printed & Coloured by R. Havell & Son, London

卡罗莱纳（长尾）鹦鹉

学　名：*Conuropsis carolinensis*

英文名：Carolina parakeet

1. 雄性　2. 雌性　3. 亚成体

Red headed Woodpecker *Male 1. F. 2. Young 3, 4, 5.*

Drawn from Nature & Published by John J. Audubon F.R.S.E. M.W.S. PICUS ERYTHROCEPHALUS Engraved, Printed & Coloured by R. Havell & Son, London

红头啄木鸟

学　名：*Melanerpes erythrocephalus*

英文名：Red-headed Woodpecker

1. 雄性　2. 雌性　3,4,5. 亚成体

Vireo Solitarius Male 1. F. 2.

SOLITARY FLYCATCHER

Plant, Vulgo Cane

Drawn from Nature and Published by John J. Audubon F.R.S.E. M.W.S.

Engraved, Printed & Coloured by R. Havell & Son, London

蓝头莺雀

学　名：*Vireo solitarius*

英文名：Blue-headed Vireo

1. 雄性　2. 雌性

Towee Bunting Male 1. F. 2.

FRINGILLA ERYTHROPTHALMA

Drawn from Nature and Published by John J. Audubon F.R.S.E. M.W.S.　*Plant, Vulgo, Black-berry*　Engraved, Printed and Coloured by R. Havell & Son, London

棕胁唧鹀

学　名：*Pipilo erythrophthalmus*

英文名：Eastern Towhee

1．雄性　2．雌性

Vigors Vireo Male

VIREO VIGORSII

Drawn from Nature and Published by John J. Audubon F.R.S.E. M.W.S. *Plant, Tradescantia Virginica* Engraved, Printed & Coloured by R. Havell & Son, London

松　莺（雄性）

学　名：*Sefophaga pinus*

（原为 *Dendroica* 属）

英文名：Pine Warbler

White-headed Eagle Male

FALCO LEUCOCEPHALUS

Fish Vulgo—Yellow mud Cat

Drawn from Nature & Published by John J. Audubon F.R.S.E. F.L.S. M.W.S.

Engraved, Printed & Coloured by R. Havell & Son, London 1828

白头海雕（雄性）

学　名：*Haliaeetus leucocephalus*

英文名：Bald Eagle

Black-billed Cuckoo Male 1. F. 2.

COCCYZUS ERYTHROPHTHALMUS

Plant, Magnolia grandiflora

Drawn from Nature and Published by
John J. Audubon F.R.S.E. F.L.S. M.W.S.

Engraved, Printed & Coloured by R. Havell & Son, London 1828

黑嘴美洲鹃

学　名：*Coccyzus erythropthalmus*

英文名：Black-billed Cuckoo

1．雄性　2．雌性

No. 7. PLATE 33.

Yellow Bird or American Goldfinch

CARDUELIS AMERICANA *Male 1. F. 2.*

Plant, Cnicus lanceolatus, Vulgo Common Thistle

Drawn from Nature and Published by John J. Audubon F.R.S.E. F.L.S. M.W.S. Engraved, Printed & Coloured by R. Havell & Son, London 1828

美洲金翅雀

学　名：*Spinus tristis*

（原为 *Carduelis* 属）

英文名：American Goldfinch

1. 雄性　2. 雌性

Worm-eating Warbler Male 1. F. 2.

DACNIS VERMIVORA

Plant Phytolacca decandra. Vulgo, Poke-berry

Drawn from Nature and Published by John J. Audubon F.R.S.E. F.L.S. M.W.S.

Engraved, Printed & Coloured by R. Havell & Son, London 1828

食虫莺

学　名：*Helmitheros vermivorum*

英文名：Worm-eating Warbler

1. 雄性　2. 雌性

黄林莺

学　名：*Setophaga petechia*

　　　（原为 *Dendroica* 属）

英文名：Mangrove Warbler

1. 雄性　2. 雌性

Stanley Hawk Male 1. F. 2.

Drawn from Nature & Published by John J. Audubon F.R.S.E. F.L.S. M.W.S. ASTUR STANLEII Engraved, Printed & Coloured by R. Havell & Son, London 1828

库氏鹰

学　名：*Accipiter cooperii*

英文名：Cooper's Hawk

1. 雄性　2. 雌性

Gold-winged Woodpecker Male 1. F. 2.

PICUS AURATUS

Drawn from Nature and Published by John J. Audubon F.R.S.E. F.L.S. M.W.S.

Engraved by Robt. Havell Junr. Printed & Coloured by R. Havell Senr. London 1828

北扑翅䴕

学　名：*Colaptes auratus*

英文名：Northern Flicker

1. 雄性　2. 雌性

Kentucky Warbler Male 1. F. 2.

SYLVIA FORMOSA

Drawn from Nature and Published by John J. Audubon F.R.S.E. F.L.S. M.W.S. *Plant Magnolia auriculata* Engraved, Printed & Coloured by R. Havell & Son, London 1828

黄腹地莺

学　名：*Geothlypis formosus*

（原为 *Oporornis* 属）

英文名：Kentucky Warbler

1. 雄性　2. 雌性

Crested Titmouse Male 1. F. 2.

PARUS BICOLOR

Drawn from Nature and Published by John J. Audubon F.R.S.E. F.L.S. M.W.S.

Plant, Pinus Strobus

Engraved, Printed & Coloured by R. Havell & Son, London 1828

美洲凤头山雀

学　名：*Baeolophus bicolor*

（原为 *Parus* 属）

英文名：Tufted Titmouse

1. 雄性　2. 雌性

American Redstart Male 1. F. 2.

MUSCICAPA RUTICILLA

Plant Vulgo, Scrub Elm

Drawn from Nature and Published by John J. Audubon F.R.S.E. F.L.S. M.W.S. Ostrya Virginica Engraved by Robt. Havell Junr. Printed & Coloured by R. Havell Senr. London 1828

橙尾鸲莺

学　名：*Setophaga ruticilla*

英文名：American Redstart

1. 雄性　2. 雌性

Ruffed Grous *Male 1 & 2. F. 3.*

Drawn from Nature & Published by John J. Audubon F.R.S.E. F.L.S. M.W.S.

TETRAO UMBELLUS *Vulgo. Pheasant*

Engraved by R. Havell Junr. Printed & Coloured by R. Havell Senr. London 1828

披肩榛鸡

学　名：*Bonasa umbellus*

英文名：Ruffed Grouse

1,2.　雄性　3.　雌性

Orchard Oriole

ICTERUS SPURIUS

Plant Vulgo, honey Locust

Gleditschia triacanthos

Male 1, 2. adult
3&4. 2nd and 3rd Year
5. Female

Drawn from Nature and Published by John J. Audubon F.R.S.E. F.L.S. M.W.S.

Engraved by R. Havell Junr. Printed & Coloured by R. Havell Senr. London 1828

圃拟鹂

学　名：*Icterus spurius*

英文名：Orchard Oriole

1,2. 成年雄性　3,4. 分别为2岁、3岁的雄性　5. 雌性

Cedar Bird Male 1. F. 2.

BOMBYCILLA CAROLINENSIS

Plant Vulgo, Red Cedar

Drawn from Nature and Published by John J. Audubon F.R.S.E. F.L.S. M.W.S. Juniper Virginiana Engraved by R. Havell Junr. Printed & Coloured by R. Havell Senr. London 1828

雪松太平鸟

学　名：*Bombycilla cedrorum*

英文名：Cedar Waxwing

1. 雄性　2. 雌性

Summer Red Bird *Male old 1. Young 2. F. 3.*

TANAGRA ÆSTIVA

Plant Vulgo, Wild Muscadine

Drawn from Nature & Published by John J. Audubon F.R.S.E. F.L.S. M.W.S. Vitis rotundifolia Engraved by R. Havell Junr. Printed & Coloured by R. Havell Senr. London 1828

玫红丽唐纳雀

学　名：*Piranga rubra*

英文名：Summer Tanager

1. 老年雄性　2. 雄性，亚成体　3. 雌性

Traill's Fly-catcher

MUSCICAPA TRAILLI

Plant Vulgo, Sweet Gum

Drawn from Nature & Published by John J. Audubon F.R.S.E. F.L.S. M.W.S.　Liquidamber Styraciflua　Engraved by R. Havell Junr. Printed & Coloured by R. Havell Senr. London 1828

桤木纹霸鹟

学　名：*Empidonax alnorum*

英文名：Alder Flycatcher

横斑林鸮（成年雄性）

学 名：*Strix varia*

英文名：Barred Owl

东部北美灰松鼠

学 名：*Sciurus carolinensis*

英文名：Eastern Squirrel

Ruby-throated Humming Bird Male 1. F. 2. Young 3.

TROCHILUS COLUBRIS

Plant, Bignania radicans

Drawn from Nature and Published by John J. Audubon F.R.S.E. F.L.S. M.W.S. Vulgo, Trumpet Flower Engraved by R. Havell Junr. Printed & Coloured by R. Havell Senr. London 1828

红喉北蜂鸟

学　名：*Archilochus colubris*

英文名：Ruby-throated Hummingbird

1. 雄性　2. 雌性　3. 亚成体

Cerulean Warbler Male 1. F. 2.

SYLVIA AZUREA

Plant Vulgo. Bear-berry

Drawn from Nature and Published by John J. Audubon F.R.S.E. F.L.S. M.W.S. Ilex Dahon Engraved by R. Havell Junr. Printed & Coloured by R. Havell Senr. London 1828

白喉林莺

学　名：*Setophaga cerulea*

（原为 *Dendroica* 属）

英文名：Cerulean Warbler

1. 雄性　2. 雌性

Blue-Green Warbler Male

SYLVIA RARA

Plant Vulgo. Spanish Mulberry

Callicarpa Americana

Drawn from Nature and Published by John J. Audubon F.R.S.E. F.L.S. M.W.S. Engraved by R. Havell Junr. Printed & Coloured by R. Havell Senr. London 1828

白喉林莺（雄性）

学　名：*Setophaga cerulea*

（原为 *Dendroica* 属）

英文名：Cerulean Warbler

Swainson's Warbler Male

SYLVICOLA SWAINSONIA

Drawn from Nature and Published by John J. Audubon F.R.S.E. F.L.S. M.W.S. *Tree. Vulgo White Oak* Engraved by R. Havell Junr. Printed & Coloured by R. Havell Senr. London 1828

纹胸林莺（雄性）

学　名：*Setophage magnolia*

（原为 *Dendroica* 属）

英文名：Magnolia Warbler

Red Tailed Hawk Male 1. F. 2.

Drawn from Nature & Published by John J. Audubon F.R.S.E. F.L.S. M.W.S. FALCO BOREALIS Engraved by R. Havell Junr. Printed & Coloured by R. Havell Senr. London 1828

红尾鵟

学　名：*Buteo jamaicensis*

英文名：Red-tailed Hawk

1. 雄性　2. 雌性

Chuck will's widow Male 1. F. 2.

CAPRIMULGUS CAROLINENSIS

Plant, Bignonia Capreolata

Drawn from Nature and Published by John J. Audubon F.R.S.E. F.L.S. M.W.S.

Engraved by R. Havell Junr. Printed & Coloured by R. Havell Senr. London–1829

卡氏夜鹰

学　名：*Antrostomus carolinensis*

（原为 *Caprimulgus* 属）

英文名：Chuck-will’s-widow

1. 雄性　2. 雌性

Painted Bunting

FRINGILLA CIRIS

1 & 2. Old Males 3. M. of 1st. Year 4. 2nd. Year 5. Female

Drawn from Nature & Published by John J. Audubon F.R.S.E. F.L.S. M.W.S. *Plant. Prunus Chicasa* Engraved by R. Havell Junr. Printed & Coloured by R. Havell Senr. London 1829

丽彩鹀

学　名：*Passerina ciris*

英文名：Painted Bunting

1,2. 老年雄性　3. 1岁雄性　4. 2岁雄性　5. 雌性

Rice Bunting Male 1. F. 2.

ICTERUS AGRIPENNIS

Drawn from Nature & Published by John J. Audubon F.R.S.E. F.L.S. M.W.S. *Plant Acer rubrum* Engraved by R. Havell Junr. Printed & Coloured by R. Havell Senr. London 1829

刺歌雀

学　名：*Dolichonyx oryzivorus*

英文名：Bobolink

1. 雄性　2. 雌性

Cuvier's Wren Male

REGULUS CUVIERI

Drawn from Nature & Published by John J. Audubon F.R.S.E. F.L.S. M.W.S. *Plant Kalmia Latifolia* Engraved by R. Havell Junr. Printed & Coloured by R. Havell Senr. London 1829

红冠戴菊（雄性）

学　名：*Regulus calendula*

英文名：Ruby-crowned Kinglet

Drawn from Nature & Published by John J. Audubon F.R.S.E. F.L.S. M.W.S. FALCO LINEATUS Engraved by R. Havell Junr. Printed & Coloured by R. Havell Senr. London 1829

赤肩鵟

学　名：*Buteo lineatus*

英文名：Red-shouldered Hawk

1. 雄性　2. 雌性

Loggerhead Shrike

LANIUS CAROLINENSIS

Drawn from Nature and Published by John J. Audubon F.R.S.E. F.L.S. M.W.S. *Plant Smilax rotundifolia* Engraved by R. Havell Junr. Printed & Coloured by R. Havell Senr. London–1829

呆头伯劳

学　名：*Lanius ludovicianus*

英文名：Loggerhead Shrike

Hermit Thrush Male 1. F. 2.

Drawn from Nature and Published by John J. Audubon F.R.S.E. F.L.S. M.W.S. TURDUS SOLITARIUS Engraved by R. Havell Junr. Printed & Coloured by R. Havell Senr. London 1829

隐夜鸫

学　名：*Catharus guttatus*

英文名：Hermit Thrush

1. 雄性　2. 雌性

Chesnut Sided Warbler Male 1. F. 2.

SYLVIA ICTEROCEPHALA

Plant Verbascum Blattaria var.
flore albicante
White-flowered Moth Mullein

Drawn from Nature and Published by John J. Audubon F.R.S.E. F.L.S. M.W.S.

Engraved by R. Havell Junr. Printed & Coloured by R. Havell Senr. London 1829

栗胁林莺

学　名：*Setophaga pensylvanica*
（原为 *Dendroica* 属）

英文名：Chestnut-sided Warbler

1. 雄性　2. 雌性

Carbonated Warbler Male 1. Young 2.

SYLVIA CARBONATA

Plant Pyrus Botryapium

Service Tree

Drawn from Nature and Published by John J. Audubon F.R.S.E. F.L.S. M.W.S.

Engraved by R. Havell Junr. Printed & Coloured by R. Havell Senr. London 1829

校注：经鉴定，奥杜邦所绘的这种鸟可能并不存在，疑似“臆想”或“创造”出来的，亦有可能在绘制某种鸟的时候特征上发生了错误。

Great horned-Owl Male 1. F. young 2.

Drawn from Nature & Published by John J. Audubon F.R.S.E. F.L.S. M.W.S. Strix Virginiana Engraved by R. Havell Junr. Printed & Coloured by R. Havell Senr. London 1829

美洲雕鸮

学　名：*Bubo virginianus*

英文名：Great Horned Owl

1．雄性　2．雌性，亚成体

No. 13. PLATE 62.

Passenger Pigeon Male 1. F. 2.

Drawn from Nature & Published by John J. Audubon F.R.S.E. F.L.S. M.W.S. COLUMBA MIGRATORIA Engraved by R. Havell Junr. Printed & Coloured by R. Havell Senr. London–1829

旅 鸽（已灭绝）

学 名：*Ectopistes migratorius*

英文名：Passenger Pigeon

1. 雄性 2. 雌性

White Eyed Flycatcher Male

VIREO NOVEBORACENSIS

Plant Melia Azedarach

Vulgo Pride of China

Drawn from Nature & Published by John J. Audubon F.R.S.E. F.L.S. M.W.S.

Engraved by R. Havell Junr. Printed & Coloured by R. Havell Senr. London 1829

白眼莺雀（雄性）

学　名：*Vireo griseus*

英文名：White-eyed Vireo

Swamp Sparrow Male

SPIZA PALUSTRIS

Plant Vulgo May Apple

Podophyllum peltatum

Drawn from Nature by Lucy Audubon

Engraved by R. Havell Junr. Printed & Coloured by R. Havell Senr. London 1829

沼泽带鹀（雄性）

学　名：*Melospiza georgiana*

英文名：Swamp Sparrow

Rathbone's Warbler *Males*

SYLVIA RATHBONI

Plant Bignonia Capreolata

Drawn from Nature and Published by John J. Audubon F.R.S.E. F.L.S. M.W.S. Engraved by R. Havell Junr. Printed & Coloured by R. Havell Senr. London 1829

黄林莺（雄性）

学　名：*Setophaga petechia*

（原为 *Dendroica* 属）

英文名：Mangrove Warbler

Ivory-billed Woodpecker Male 1. F. 2&3.

Drawn from Nature & Published by John J. Audubon F.R.S.E. F.L.S. M.W.S. PICUS PRINCIPALIS Engraved by R. Havell Junr. Printed & Coloured by R. Havell Senr. London 1829

象牙喙啄木鸟

学　名：*Campephilus principalis*

英文名：Ivory-billed Woodpecker

1. 雄性　2,3. 雌性

Red-winged Starling *Adult Male 1. Young Male 2. Female Old 3. Young 4.*

ICTERUS PHŒNICEUS

Plant Acer rubrum

Vulgo Swamp Maple

Drawn from Nature and Published by John J. Audubon F.R.S.E. F.L.S. M.W.S.

Engraved by R. Havell Junr. Printed & Coloured by R. Havell Senr. London–1829

红翅黑鹂

学　名：*Agelaius phoeniceus*

英文名：Red-winged Blackbird

1. 成年雄性　2. 雄性，亚成体　3. 老年雌性　4. 亚成体

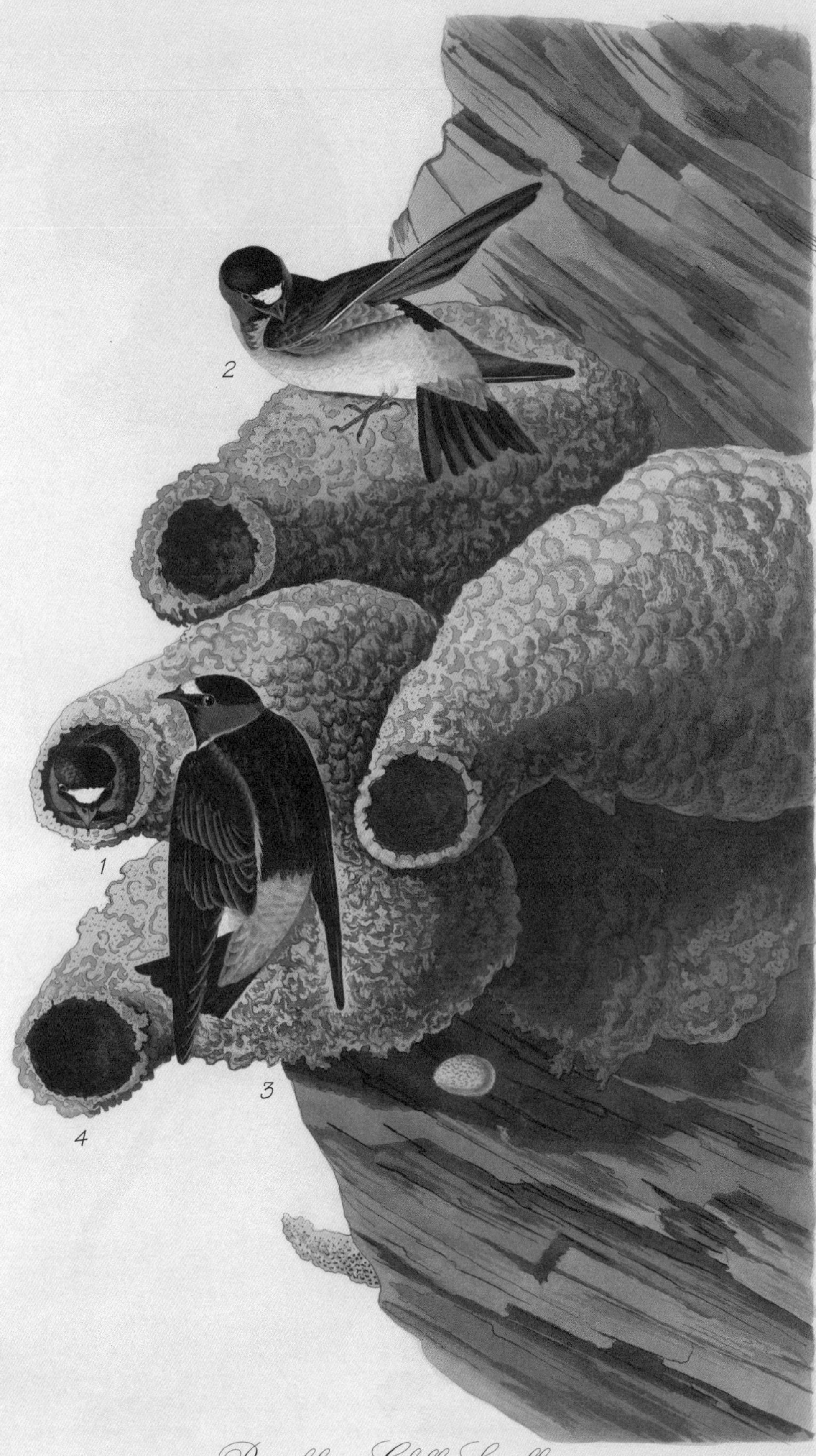

Republican Cliff Swallow Male 1. F. 2. Egg 3. Nests 4.

Drawn from Nature and Published by John J. Audubon F.R.S.E. F.L.S. M.W.S. HIRUNDO FULVA Engraved by R. Havell Junr. Printed & Coloured by R. Havell Senr. London 1829

美洲燕

学　名：*Petrochelidon pyrrhonota*

英文名：American Cliff Swallow

1. 雄性　2. 雌性　3. 卵　4. 巢

Bay breasted Warbler Male 1. F. 2.

SYLVIA CASTANEA

Plant Vulgo, Highland Cotton

Drawn from Nature & Published by John J. Audubon F.R.S.E. F.L.S. M.W.S. *Gofsipium herbaceum* Engraved by R. Havell Junr. Printed & Coloured by R. Havell Senr. London 1829

栗胸林莺

学　名：*Setophaga castanea*

（原为 *Dendroica* 属）

英文名：Bay-breasted Warbler

1．雄性　2．雌性

No. 14. PLATE 70.

Henslow's Bunting

AMMODRAMUS HENSLOWII

Spigelia Marilandica

Phlox aristata

Drawn from Nature & Published by John J. Audubon F.R.S.E. F.L.S. M.W.S. Engraved by R. Havell Junr. Printed & Coloured by R. Havell Senr. London 1829

亨氏草鹀

学　名：*Ammodramus henslowii*

英文名：Henslow's Sparrow

Winter Hawk Male Adult

CIRCUS HYEMALIS

Bull Frog

Drawn from Nature & Published by John J. Audubon F.R.S.E. F.L.S. M.W.S.

Engraved by R. Havell Junr. Printed & Coloured by R. Havell Senr. London 1829

赤肩鵟（成年雄性）

学　名：*Buteo lineatus*

英文名：Red-shouldered Hawk

No. 15.

PLATE 72.

Swallow-tailed Hawk *Male 1. Female the same*

FALCO FURCATUS

Reptile Vulgo Garter Snake

Drawn from Nature and Published by John J. Audubon F.R.S.E. F.L.S. M.W.S.

Engraved by R. Havell Junr. Printed & Coloured by R. Havell Senr. London–1829

燕尾鸢（雄性，雌雄同色）

学　名：*Elanoides forficatus*

Wood Thrush Male 1. F. 2.

TURDUS MUSTELINUS

Plant Cornus Florida

Vulgo Dog-wood

Drawn from Nature and Published by John J. Audubon F.R.S.E. F.L.S. M.W.S.

Engraved by R. Havell Junr. Printed & Coloured by R. Havell Senr. London 1829

棕林鸫

学　名：*Hylocichla mustelina*

英文名：Wood Thrush

1．雄性　2．雌性

No. 15. PLATE 74.

Indigo-bird

FRINGILLA CYANEA

Male Adult 1. M. First Year 2. 2nd. 3. F. 4   Plant Schisandra Coccinea

Drawn from Nature and Published by John J. Audubon F.R.S.E. F.L.S. M.W.S.   *Vulgo Salsparilla*   Engraved by R. Havell Junr. Printed & Coloured by R. Havell Senr. London 1829

靛彩鹀

学　名：*Passerina cyanea*

英文名：Indigo Bunting

1. 成年雄性　2. 1岁雄性　3. 2岁雄性　4. 雌性

Le petit Caporal Male

Drawn from Nature and Published by John J. Audubon F.R.S.E. F.L.S. M.W.S.   FALCO TEMERARIUS   Engraved by R. Havell Junr. Printed & Coloured by R. Havell Senr. London 1829

灰背隼（雄性）

学　名：*Falco columbarius*

英文名：Merlin

No. 16.

PLATE 76.

Virginian Partridge

Male adult 1. Young 2. Female adult 3. Young 4. very young Birds 5.

Drawn from Nature & Published by John J. Audubon F.R.S. L&E. F.L.S. &C.

PERDIX VIRGINIANA

Engraved, Printed & Coloured by R. Havell Junr. 1830

山齿鹑

学　名：*Buteo lineatus*

英文名：Red-shouldered Hawk

1. 成年雄性　2. 雄性，亚成体　3. 成年雌性　4. 雌性，亚成体　5. 雏鸟　　（注：图中6为赤肩鵟）

Belted Kingfisher Male 1& 2. F. 3.

Drawn from Nature and Published by John J. Audubon F.R.S. L&E. F.L.S. &C. ALCEDO ALCYON Engraved, Printed & Coloured by R. Havell Junr. 1830

白腹鱼狗

学　名：*Megaceryle alcyon*

（原为 *Ceryle* 属）

英文名：Belted Kingfisher

1,2. 雄性　3. 雌性

Great Carolina Wren Male 1. F. 2.

TROGLODYTES LUDOVICIANUS

Plant Vulgo Dwarf horse Chesnut

Æsculus Pavia

Drawn from Nature and Published by John J. Audubon F.R.S.E. F.L.S. M.W.S.

Engraved, Printed & Coloured by R. Havell Junr. 1830

卡罗苇鹪鹩

学　名：*Thryothorus ludovicianus*

英文名：Carolina Wren

1. 雄性　2. 雌性

Tyrant Flycatcher Male 1. F. 2.

MUSCICAPA TYRANNUS

Plant Cotton Wood

Populus candicans

Drawn from Nature and Published by John J. Audubon F.R.S. L&E. F.L.S. &C.

Engraved, Printed & Coloured by R. Havell Junr. 1830

东王霸鹟

学　名：*Tyrannus tyrannus*

英文名：Eastern Kingbird

1．雄性　2．雌性

No. 16. PLATE 80.

Anthus Hypogaeus

Drawn from Nature and Published by John J. Audubon F.R.S. L&E. F.L.S. &C.

PHLOX SUBULATA

Engraved, Printed & Coloured by R. Havell Junr. 1830

黄腹鹨

学　名：*Anthus rubescens*

英文名：Buff-bellied Pipit

Fish Hawk Male

FALCO HALIÆTUS

Vulgo Weak Fish

Drawn from Nature & Published by John J. Audubon F.R.S. F.L.S. &C.

Engraved, Printed & Coloured by R. Havell Junr. 1830

鹗（雄性）

学　名：*Pandion haliaetus*

英文名：Western Osprey

Caprimulgus vociferus

Whip-poor-will Male 1. F. 2, 3.

quercus tinctoria

Drawn from Nature and Published by John J. Audubon F.R.S. F.L.S. &C. *Vulgo Black Oak* Engraved, Printed & Coloured by R. Havell Junr. 1830

三声夜鹰

学　名：*Caprimulgus vociferus*

英文名：Mexican Whip-poor-will

1. 雄性　2,3. 雌性

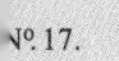

House Wren Male 1. F. 2. Young 3, 4, 5.

Drawn from Nature and Published by John J. Audubon F.R.S. F.L.S. &C. TROGLODYTES ÆDON Engraved, Printed & Coloured by R. Havell Junr. 1830

莺鹪鹩

学　名：*Troglodytes aedon*

英文名：House Wren

1. 雄性　2. 雌性　3,4,5. 亚成体

Blue Grey Flycatcher Male 1. F. 2.

SYLVIA CŒRULA

Plant Juglans nigra
Vulgo Black Walnut

Drawn from Nature and Published by John J. Audubon F.R.S. F.L.S. &C.

Engraved, Printed & Coloured by R. Havell Junr. 1830

灰蓝蚋莺

学　名：*Polioptila caerulea*

英文名：Blue-gray Gnatcatcher

1. 雄性　2. 雌性

Yellow Throat Warbler Male

SYLVIA PENSILIS

Plant Castanea pumila

Vulgo Chink-apin

Drawn from Nature & Published by John J. Audubon F.R.S. F.L.S. &C..

Engraved, Printed & Coloured by R. Havell Junr. 1830

黄喉林莺（雄性）

学　名：*Setophaga dominica*

（原为 *Dendroica* 属）

英文名：Yellow-throated Warbler

红尾鵟

学　名：*Buteo jamaicensis*

英文名：Red-tailed Hawk

1. 雄性　2. 雌性

Florida Jay *Male 1. F. 2.*

GARRULUS FLORIDANUS

Diospyros Virginiana

Vulgo Persimon

Drawn from Nature and Published by John J. Audubon F.R.S's. L&E. F.L.S. &C.

Engraved, Printed & Coloured by R. Havell Junr. London–1830

丛 鸦

学　名：*Aphelocoma coerulescens*

英文名：Florida Scrub Jay

1. 雄性　2. 雌性

Autumnal Warbler Male 1. F. 2.

SYLVIA AUTUMNALIS

Plant Betula papyrifera

Vulgo Canœ Birch

Drawn from Nature and Published by John J. Audubon F.R.S's. L&E. F.L.S. &C.

Engraved, Printed & Coloured by R. Havell Junr. London 1830

栗胸林莺

学　名：*Setophaga castanea*

（原为 *Dendroica* 属）

英文名：Bay-breasted Warbler

1. 雄性　2. 雌性

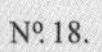

PLATE 89.

Nashville Warbler *Male 1. F. 2.*

SYLVIA RUBRICAPILLA

Plant Ilex

Vulgo Spice Wood

Drawn from Nature and Published by John J. Audubon F.R.S's. L&E. F.L.S. &C.

Engraved, Printed & Coloured by R. Havell Junr. London 1830

黄喉史森莺

学　名：*Leiothlypis ruficapilla*

（原为 *Vermivora* 属）

英文名：Nashville Warbler

1. 雄性　2. 雌性

Black and white Creeper

SYLVIA VARIA

Pinus pendula

Vulgo Black Larch

Drawn from Nature and Published by John J. Audubon F.R.S's. L&E. F.L.S. &C.

Engraved, Printed & Coloured by R. Havell Junr. London 1830

黑白森莺

学　名：*Mniotilta varia*

英文名：Black-and-white Warbler

Broad-winged Hawk *Male 1. F. 2.*

FALCO PENNSYLVANICUS

Plant Juglans porcina

Vulgo Pig-nut

Drawn from Nature & Published by John J. Audubon F.R.S's. L&E. F.L.S.&C.

Engraved, Printed & Coloured by R. Havell Junr. London 1830

巨翅鵟

学　名：*Buteo platypterus*

英文名：Broad-winged Hawk

1. 雄性　2. 雌性

Pigeon Hawk Male 1. F. 2.

Drawn from Nature and Published by John J. Audubon F.R.S's. L&E. F.L.S.&C. FALCO COLUMBARIUS Engraved, Printed & Coloured by R. Havell Junr. London–1830

灰背隼

学　名：*Falco columbarius*

英文名：Merlin

1. 雄性　2. 雌性

Sea-side Finch

FRINGILLA MARITIMA

Plant Rosa Carolina

Vulgo Wild Rose

Drawn from Nature and Published by John J. Audubon F.R.S's. L&E. F.L.S. &C.

Engraved, Printed & Coloured by R. Havell Junr. London 1830

海滨沙鹀

学　名：*Ammodramus maritimus*

英文名：Seaside Sparrow

Bay-winged Bunting Male

FRINGILLA GRAMINEA

Plant Cactus opuntia

Vulgo Prickly Pear

Drawn from Nature and Published by John J. Audubon F.R.S's. L&E. F.L.S. &C.

Engraved, Printed & Coloured by R. Havell Junr. London 1830

栗肩雀鹀（雄性）

学　名：*Pooecetes gramineus*

英文名：Vesper Sparrow

Blue-eyed yellow Warbler

SYLVIA ÆSTIVA

Drawn from Nature and Published by John J. Audubon F.R.S's. L&E. F.L.S. &C. *Plant Wisterea* Engraved, Printed & Coloured by R. Havell Junr. London 1830

黄林莺

学　名：*Setophaga petechia*

（原为 *Dendroica* 属）

英文名：Mangrove Warbler

Columbia Jay Male 1. F. 2.

Drawn from Nature & Published by John J. Audubon F.R.S's. L&E. F.L.S. &C.　GARRULUS ULTRAMARINUS　Engraved, Printed & Coloured by R. Havell Junr. London 1830

白喉鹊鸦

学　名：*Calocitta formosa*

英文名：White-throated Magpie-jay

1．雄性　2．雌性

Mottled Owl *Adult 1. Young 2&3.*

STRIX ASIO

Plant Pinus inops

Vulgo Jersey Pine

Drawn from Nature and Published by John J. Audubon F.R.S's. L&E. F.L.S. &C.

Engraved, Printed & Coloured by R. Havell Junr. London–1830

东美角鸮

学　名：*Megascops asio*

（原为 *Otus* 属）

英文名：Eastern Screech Owl

1．成体　2,3．亚成体

长嘴沼泽鹪鹩

学　名：*Cistothorus palustris*

英文名：Marsh Wren

1. 雄性　2,3. 雌性　4. 巢

Cow Bunting Male 1. F. 2.

Drawn from Nature and Published by John J. Audubon F.R.S's. L&E. F.L.S. &C.

ICTERUS PECORIS

Engraved, Printed & Coloured by R. Havell Junr. London 1830

褐头牛鹂

学　名：*Molothrus ater*

英文名：Brown-headed Cowbird

1．雄性　2．雌性

Green-blue, or White, Bellied Swallow Male 1. F. 2.

Drawn from Nature and Published by John J. Audubon F.R.S's. L&E. F.L.S. &C. HIRUNDO BICOLOR Engraved, Printed & Coloured by R. Havell Junr. London 1830

双色树燕

学　名：*Tachycineta bicolor*

英文名：Tree Swallow

1. 雄性　2. 雌性

Raven Male

CORVUS CORAX

Drawn from Nature by J. J. Audubon F.R.S. F.L.S. *Thick Shell-bark Hickory. Juglans laciniosa* Engraved, printed & Coloured by R. Havell, Junr.

渡　鸦（雄性）

学　名：*Corvus corax*

英文名：Northern Raven

Blue Jay *Male 1. Female 2,3.*

Drawn from Nature by J. J. Audubon F.R.S. F.L.S. CORVUS CRISTATUS Engraved, printed & Coloured by R. Havell, Junr

冠蓝鸦

学　名：*Cyanocitta cristata*

英文名：Blue Jay

1. 雄性　2,3. 雌性

No.21. PLATE CIII.

SYLVIA PARDALINA

Drawn from Nature by J. J. Audubon F.R.S. F.L.S. *Big Laurel. Rhododendron Maximum* Engraved, printed & Coloured by R. Havell, Junr.

加拿大威森莺

学　名：*Cardellina canadensis*

（原为 *Wilsonia* 属）

英文名：Canada Warbler

1. 雄性　2. 雌性

Chipping Sparrow Male

FRINGILLA SOCIALIS

Drawn from Nature by J. J. Audubon F.R.S. F.L.S.

Black Locust. Robinia Pseudacacia

Engraved, printed & Coloured by R. Havell, Junr.

棕顶雀鹀（雄性）

学　名：*Spizella passerina*

英文名：Chipping Sparrow

Red-breasted Nuthatch Male 1. Female 2.

Drawn from Nature by J. J. Audubon F.R.S. F.L.S. SITTA CANADENSIS Engraved, printed & Coloured by R. Havell, Junr.

红胸䴓

学 名：*Sitta canadensis*

英文名：Red-breasted Nuthatch

1. 雄性 2. 雌性

No. 22. PLATE CVI.

Black Vulture or Carrion Crow *Male 1. and Female 2.* CATHARTES ATRATUS *American Deer. Cervus virginianus*

Drawn from Nature by J. J. Audubon F.R.S. F.L.S.

Engraved, Printed & Coloured by R. Havell, London 1831

黑头美洲鹫

学　名：*Coragyps atratus*

英文名：Black Vulture

1. 雄性　2. 雌性

Canada Jay

CORVUS CANADENSIS, Linn.

Male 1. Female 2.

White Oak. Quercus alba

Drawn from Nature by J. J. Audubon F.R.S. F.L.S.

Engraved, Printed & Coloured by R. Havell, London 1831

灰噪鸦

学　名：*Perisoreus canadensis*

英文名：Grey Jay

1. 雄性　2. 雌性

No. 22.

PLATE CVIII.

Fox-coloured Sparrow

FRINGILLA ILIACA

Male 1. Female 2.

Drawn from Nature by J. J. Audubon F.R.S. F.L.S.

Engraved, Printed & Coloured by R. Havell, London 1831

狐色雀鹀

学　名：*Passerella iliaca*

英文名：Red Fox Sparrow

1. 雄性　2. 雌性

Savannah Finch FRINGILLA SAVANNA *Male 1. Female 2.*

Drawn from Nature by J. J. Audubon F.R.S. F.L.S.　Engraved, Printed & Coloured by R. Havell, London 1831

稀树草鹀

学　名：*Passerculus sandwichensis*

英文名：Savannah Sparrow

1. 雄性　2. 雌性

黑枕威森莺

学　名：*Setophaga citrina*

（原为 *Wilsonia* 属）

英文名：Hooded Warbler

1. 雄性　2. 雌性

No. 23. PLATE CXI.

Pileated Woodpecker

PICUS PILEATUS, Linn.

Adult Male 1. Adult Female 2. Young Males 3, 4.

Racoon Grape. Vitis astivalis

Drawn from Nature by J. J. Audubon F.R.S. F.L.S.

Engraved, Printed & Coloured by R. Havell

北美黑啄木鸟

学　名：*Dryocopus pileatus*

英文名：Pileated Woodpecker

1. 成年雄性　2. 成年雌性　3,4. 雄性，亚成体

Downy Woodpecker

PICUS PUBESCENS

Male 1. Female 2.

Drawn from Nature by J. J. Audubon F.R.S. F.L.S..

Bignonia capreolata

Engraved, Printed & Coloured by R. Havell, London. 1831

绒啄木鸟

学　名：*Picoides pubescens*

英文名：Downy Woodpecker

1. 雄性　2. 雌性

Blue-bird

SYLVIA SIALIS

Male 1. Female 2. Young 3.

Great Mullein Verbascum Thapsus

Drawn from Nature by J. J. Audubon F.R.S. F.L.S.

Engraved, Printed & Coloured by R. Havell, London 1831

东蓝鸲

学　名：*Sialia sialis*

英文名：Eastern Bluebird

1. 雄性　2. 雌性　3. 亚成体

White-crowned Sparrow

FRINGILLA LEUCOPHRYS

Male 1. Female 2.

Summer Grape. Vitis Astivalis

Drawn from Nature by J. J. Audubon F.R.S. F.L.S.

Engraved, Printed & Coloured by R. Havell, London 183

白冠带鹀

学　名：*Zonotrichia leucophrys*

英文名：White-crowned Sparrow

1. 雄性　2. 雌性

Wood Pewee MUSCICAPA VIRENs *Male* *Swamp Honeysuckle. Azalea biscosa*

Drawn from Nature by J. J. Audubon F.R.S. F.L.S.

Engraved, Printed & Coloured by R. Havell, London 1831

东绿霸鹟（雄性）

学　名：*Contopus virens*

英文名：Eastern Wood Pewee

Ferruginous Thrush

TURDUS RUFUS, Linn.

Male 1. Female 2.

Black-jack Oak, Quercus nigra

Black Snake

Drawn from Nature by J. J. Audubon F.R.S. F.L.S.

Engraved, Printed & Coloured by R. Havell, London

褐弯嘴嘲鸫

学　名：*Toxostoma rufum*

英文名：Brown Thrasher

1. 雄性　2. 雌性

Mifsifsippi Kite

FALCO PLUMBEUS, Gmel.

Drawn from Nature by J. J. Audubon F.R.S. F.L.S. *Male 1. Female 2.* Engraved, Printed & Coloured by R. Havell.

密西西比灰鸢

学　名：*Ictinia mississippiensis*

英文名：Mississippi Kite

1. 雄性　2. 雌性

Warbling Flycatcher

MUSCICAPA GILVA, Vieill

Male 1. Female 2.

Drawn from Nature by J. J. Audubon F.R.S. F.L.S.　*Swamp Magnolia Magnolia Glauca*　Engraved, Printed & Coloured by R. Havell

歌莺雀

学　名：*Vireo gilvus*

英文名：Warbling Vireo

1．雄性　2．雌性

Yellow-throated Vireo

VIREO FLAVIFRONS, Vieill

Male

Drawn from Nature by J. J. Audubon F.R.S. F.L.S.

Swamp Snow-ball. Hydrangea quercifolia

Engraved, Printed & Coloured by R. Havell, London

黄喉莺雀（雄性）

学　名：*Vireo flavifrons*

英文名：Yellow-throated Vireo

Pewit Flycatcher

MUSCICAPA FUSCA, Gmel.

Male 1. Female 2.

Cotton plant Gossypium

Drawn from Nature by J. J. Audubon F.R.S. F.L.S.

Engraved, Printed & Coloured by R. Havell

灰胸长尾霸鹟

学　名：*Sayornis phoebe*

英文名：Eastern Phoebe

1. 雄性　2. 雌性

Drawn from Nature by J. J. Audubon F.R.S. F.L.S. *Snowy Owl* STRIX NYCTEA, Linn. *Male 1. Female 2.* Engraved, Printed & Coloured by R. Havell, London

雪 鸮

学　名：*Bubo scandiaca*

英文名：Snowy Owl

1. 雄性　2. 雌性

Blue Grosbeak FRINGILLA CORULEA, Bonap *Male 1. Female 2. Young 3. Dog-wood Cornus florida*

Drawn from Nature by J. J. Audubon F.R.S. F.L.S. Engraved, Printed & Coloured by R. Havell, London

斑翅蓝彩鹀

学　名：*Passerina caerulea*

（原为 *Guiraca* 属）

英文名：Blue Grosbeak

1. 雄性　2. 雌性　3. 亚成体

Black & Yellow Warbler SYLVIA MACULOSA, Lath. *Male 1. Female 2. Flowering Rasp-berry. Rubus odoratus*

Drawn from Nature by J. J. Audubon F.R.S. F.L.S.　　Engraved, Printed & Coloured by R. Havell, London

纹胸林莺

学　名：*Setophaga magnolia*

（原为 *Dendroica* 属）

英文名：Magnolia Warbler

1. 雄性　2. 雌性

Green Black-capt Flycatcher MUSCICAPA PUSILLA, Wils. *Male 1. Female 2. Snakes-head. Chelone glabra*

Drawn from Nature by J. J. Audubon F.R.S. F.L.S.

Engraved, Printed & Coloured by R. Havell, London

黑头威森莺

学　名：*Cardellina pusilla*

（原为 *Wilsonia* 属）

英文名：Wilson's Warbler

1. 雄性　2. 雌性

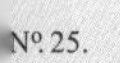

PLATE CXXV.

Brown-headed Nuthatch SITTA PUSILLA, Lath. *Male 1. Female 2.*

Drawn from Nature by J. J. Audubon F.R.S. F.L.S.

Engraved, Printed & Coloured by R. Havell, London

褐头䴓

学　名：*Sitta pusilla*

英文名：Brown-headed Nuthatch

1. 雄性　2. 雌性

White-headed Eagle

FALCO LEUCOCEPHALUS, Linn.

Drawn from Nature by J. J. Audubon F.R.S. F.L.S. *Young* Engraved, Printed & Coloured by R. Havell, London

白头海雕（亚成体）

学　名：*Haliaeetus leucocephalus*

英文名：Bald Eagle

Rose-breasted Grosbeak FRINGILLA LUDOVICIANA, Bonap *Male 1. Female 2. Young in autumn 3. Young 4. Ground Hemlock, Taxus canadensis*

Drawn from Nature by J. J. Audubon F.R.S. F.L.S.

Engraved, Printed & Coloured by R. Havell, London

玫胸白斑翅雀

学　名：*Pheucticus ludovicianus*

英文名：Rose-breasted Grosbeak

1. 雄性　2. 雌性　3. 秋季的亚成体　4. 亚成体

Cat Bird

TURDUS FELIVOX, Vieill

Male 1. Female 2.

Black-berry. Rubus villosus

Drawn from Nature by J. J. Audubon F.R.S. F.L.S.

Engraved, Printed & Coloured by R. Havell, London

灰嘲鸫

学　名：*Dumetella carolinensis*

英文名：Grey Catbird

1. 雄性　2. 雌性

Nº 26. PLATE CXXIX.

Great Crested Flycatcher

MUSCICAPA CRINITA, Linn.

Drawn from Nature by J. J. Audubon F.R.S. F.L.S. *Males* Engraved, Printed & Coloured by R. Havell

大冠蝇霸鹟（雄性）

学　名：*Myiarchus crinitus*

英文名：Great-crested Flycatcher

Yellow-winged Sparrow

Drawn from Nature by J. J. Audubon F.R.S. F.L.S. FRINGILLA PASSERINA, Wils. *Male Phlox subulata* Engraved, Printed & Coloured by R. Havell, London

黄胸草鹀（雄性）

学　名：*Ammodramus savannarum*

英文名：Grasshopper Sparrow

American Robin

TURDUS MIGRATORIUS

Male 1. Female 2. Young 3.

Chestnut oak. Quercus Prinus

Drawn from Nature by J. J. Audubon F.R.S. F.L.S.

Engraved, Printed & Coloured by R. Havell, London 1832

旅 鸫

学 名：*Turdus migratorius*

英文名：American Robin

1. 雄性 2. 雌性 3. 亚成体

No. 27.

PLATE CXXXII

Three-toed Woodpecker PICUS TRIDACTYLUS, Linn. *Males 1. Female 2.*

Drawn from Nature by J. J. Audubon F.R.S. F.L.S.

Engraved, Printed & Coloured by R. Havell, London 183

黑背啄木鸟

学　名：*Picoides arcticus*

英文名：Black-backed Woodpecker

1. 雄性　2. 雌性

Black-poll Warbler

SYLVIA STRIATA, Lath.

Male 1. Female 2.

Black Gum Tree. Nyssa aquatica

Drawn from Nature by J. J. Audubon F.R.S. F.L.S.   Engraved, Printed & Coloured by R. Havell, London 1832

白颊林莺

学　名：*Setophaga striata*

（原为 *Dendroica* 属）

英文名：Blackpoll Warbler

1. 雄性　2. 雌性

Hemlock Warbler

SYLVIA PARUS, Wils.

Male 1. Female 2.

Dwarf Maple. Acer spicatum

Drawn from Nature by J. J. Audubon F.R.S. F.L.S.

Engraved, Printed & Coloured by R. Havell, London 183

橙胸林莺

学　名：*Setophaga fusca*

（原为 *Dendroica* 属）

英文名：Blackburnian Warbler

1. 雄性　2. 雌性

Blackburnian Warbler

SYLVIA BLACKBURNIA, Lath.

Male

Phlox maculata

Drawn from Nature by J. J. Audubon F.R.S. F.L.S.

Engraved, Printed & Coloured by R. Havell, London 1832

橙胸林莺（雄性）

学　名：*Setophaga fusca*

（原为 *Dendroica* 属）

英文名：Blackburnian Warbler

Meadow Lark STURNUS LUDOVICIANUS, Linn. *Males 1. Females 2. Gerardia flava*

Drawn from Nature by J. J. Audubon F.R.S. F.L.S.

Engraved, Printed & Coloured by R. Havell, London 1832

东草地鹨

学　名：*Sturnella magna*

英文名：Eastern Meadowlark

1. 雄性　2. 雌性

Nº. 28. PLATE CXXXVII.

Yellow-breasted Chat ICTERIA VIRIDIS, Bonap *Males 1. Female 2. Sweet Briar Rosa rubiginosa*

Drawn from Nature by J. J. Audubon F.R.S. F.L.S. Engraved, Printed & Coloured by R. Havell, London 1832

黄胸大鹏莺

学　名：*Icteria virens*

英文名：Yellow-breasted Chat

1. 雄性　2. 雌性

Connecticut Warbler

SYLVIA AGILIS, Wils.

Male 1. Female 2.

Gentiana saponaria

Drawn from Nature by J. J. Audubon F.R.S. F.L.S.

Engraved, Printed & Coloured by R. Havell, London 1832

灰喉地莺

学　名：*Oporornis agilis*

英文名：Connecticut Warbler

1. 雄性　2. 雌性

Field Sparrow FRINGILLA PUSILLA, Wils.

Drawn from Nature by J. J. Audubon F.R.S. F.L.S.. *Male* *Calopogon pulchellum & Vaccinium tenellum* Engraved, Printed & Coloured by R. Havell, London 1832

田雀鹀（雄性）

学　名：*Spizella pusilla*

英文名：Field Sparrow

Pine Creeping Warbler

Drawn from Nature by J. J. Audubon F.R.S. F.L.S. SYLVIA PINUS, Lath. *Male 1. Female 2. Yellow Pine* Engraved, Printed & Coloured by R. Havell, London 1832

松 莺

学 名：*Setophaga pinus*

（原为 *Dendroica* 属）

英文名：Pine Warbler

1. 雄性 2. 雌性

Goshawk — *Stanley Hawk*

FALCO PALUMBARIUS, Linn. *Adult Male 1. Young 2.*

FALCO STANLEII, Aud. *Adult 3.*

Drawn from Nature by J. J. Audubon F.R.S. F.L.S.

Engraved, Printed & Coloured by R. Havell, London

1, 2

苍　鹰

学　名：*Accipiter gentilis*

英文名：Northern Goshawk

1. 成年雄性　2. 亚成体

3

库氏鹰（成体）

学　名：*Accipiter cooperli*

英文名：Coop's Hawk

American Sparrow Hawk FALCO SPARVERIUS, Linn. *Male 1. Female 2. Butler-nut or White walnut Juglans cinerea*

Drawn from Nature by J. J. Audubon F.R.S. F.L.S.

Engraved, Printed & Coloured by R. Havell, London

美洲隼

学　名：*Falco sparverius*

英文名：American Kestrel

1. 雄性　2. 雌性

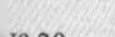

Golden-crowned Thrush

TURDUS AUROCAPILLUS, Wils. *Male 1. Female 2. Woody Nightshade. Solanum Dulcamara*

Drawn from Nature by J. J. Audubon F.R.S. F.L.S.

Engraved, Printed & Coloured by R. Havell, London 1832

橙顶灶莺

学　名：*Seiurus aurocapilla*

英文名：Ovenbird

1. 雄性　2. 雌性

No. 29. PLATE CXLIV.

Drawn from Nature by J. J. Audubon F.R.S. F.L.S.

Sassafras. Laurus Sassafras

Engraved, Printed & Coloured by R. Havell, London 1832

绿纹霸鹟

学　名：*Empidonax virescens*

英文名：Acadian Flycatcher

1．雄性　2．雌性

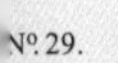

Yellow Red-poll Warbler

SYLVIA PETECHIA, Lath.

Male 1. Female 2.

Drawn from Nature by J. J. Audubon F.R.S. F.L.S.

Helenium quadridentatum

Engraved, Printed & Coloured by R. Havell, London 1832

棕榈林莺

学　名：*Setophaga palmarum*

（原为 *Dendroica* 属）

英文名：Palm Warbler

1. 雄性　2. 雌性

No. 30. PLATE CXLVI.

Fish Crow

CORVUS OSSIFRAGUS, Wils.

Male 1. Female 2.

Drawn from Nature by J. J. Audubon F.R.S. F.L.S. *Vulgo Honey–Locust. Gleditschia Hiacanthos* Engraved, Printed & Coloured by R. Havell, London 1832

鱼 鸦

学 名：*Corvus ossifragus*

英文名：Fish Crow

1. 雄性 2. 雌性

Night Hawk

CAPRIMULGUS VIRGINIANUS, Brifs

Male 1. Female 2.

White Oak. Quercus alba

Drawn from Nature by J. J. Audubon F.R.S. F.L.S.

Engraved, Printed & Coloured by R. Havell, London 1832

美洲夜鹰

学　名：*Chordeiles minor*

英文名：Common Nighthawk

1. 雄性　2. 雌性

Pine Swamp Warbler

SYLVIA SPHAGNOSA, Bonap

Male 1. Female 2.

Drawn from Nature by J. J. Audubon F.R.S. F.L.S. *Hobble Bush. Vibernum lantanoides.* Engraved, Printed & Coloured by R. Havell, London 1832

黑喉蓝林莺

学　名：*Setophaga caerulescens*

（原为 *Dendroica* 属）

英文名：Black-throated Blue Warbler

1. 雄性　2. 雌性

Sharp-tailed Finch

FRINGILLA CAUDACUTA, Wils.

Drawn from Nature by J. J. Audubon F.R.S. F.L.S. *Male 1. Female 2.* Engraved, Printed & Coloured by R. Havell, London 1832

尖尾沙鹀

学　名：*Ammodramus caudacutus*

英文名：Saltmarsh Sparrow

1. 雄性　2. 雌性

Red-eyed Vireo

VIREO OLIVACEUS, Bonap

Male

Drawn from Nature by J. J. Audubon F.R.S. F.L.S. *Honey Locust Gleditschia hiacanthos* Engraved, Printed & Coloured by R. Havell, London 1832

红眼莺雀（雄性）

学　名：*Vireo olivaceus*

英文名：Red-eyed Vireo

Turkey Buzzard

CATHARTES ATRATUS

Drawn from Nature by J. J. Audubon F.R.S. F.L.S. *Male 1. Young 2.* Engraved, Printed & Coloured by R. Havell, London 1832

红头美洲鹫

学　名：*Cathartes aura*

英文名：Turkey Vulture

1. 雄性　2. 亚成体

White-breasted Black-capped Nuthatch

Drawn from Nature by J. J. Audubon F.R.S. F.L.S. SITTA CAROLINENSIS, Brifs *Male 1. Female 2.* Engraved, Printed & Coloured by R. Havell, London 1832

白胸䴓

学　名：*Sitta carolinensis*

英文名：White-breasted Nuthatch

1. 雄性　2. 雌性

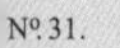

Yellow-crown Warbler

SYLVIA CORONATA, Lath.

Male 1. Young 2.

Iris versicolor

Drawn from Nature by J. J. Audubon F.R.S. F.L.S.

Engraved, Printed & Coloured by R. Havell, London 1832

黄腰林莺

学　名：*Setophaga coronata*

（原为 *Dendroica* 属）

英文名：Yellow-rumped Warbler

1. 雄性　2. 亚成体

SYLVIA PEREGRINA, Wils.

Male

Drawn from Nature by J. J. Audubon F.R.S. F.L.S. *Prunus*. sp. Engraved, Printed & Coloured by R. Havell, London 1832

灰冠虫森莺（雄性）

学　名：*Oreothlypis peregrina*

（原为 *Helmintophila, Leiothlypis, Vermivora* 属）

英文名：Tennessee Warbler

Black-throated Blue Warbler

SYLVIA CANADENSIS, Lath.

Male

Drawn from Nature by J. J. Audubon F.R.S. F.L.S. *Canadian Columbine. Aquilegia canadensis* Engraved, Printed & Coloured by R. Havell, London 1832

黑喉蓝林莺（雄性）

学　名：*Setophaga caerulescens*

（原为 *Dendroica* 属）

英文名：Black-throated Blue Warbler

American Crow

CORVUS AMERICANUS

Male

Black Walnut, Corvus americanus

Nest of the Ruby-throated Humming Bird

Drawn from Nature by J. J. Audubon F.R.S. F.L.S.

Engraved, Printed & Coloured by R. Havell, London

短嘴鸦（雄性）

学　名：*Corvus brachyrhynchos*

英文名：American Crow

Rusty Grakle

QUISCALUS FERRUGINEUS. Bonap

Male 1. Female 2. Young 3.

Black Haw

Drawn from Nature by J. J. Audubon F.R.S. F.L.S.

Engraved, Printed & Coloured by R. Havell, London 1833

锈色黑鹂

学　名：*Euphagus carolinus*

英文名：Rusty Blackbird

1. 雄性　2. 雌性　3. 亚成体

No. 32. PLATE CLVIII

American Swift

CYPSELUS PELASGIUS, Temm

Male 1. Female 2.

Nests

Drawn from Nature by J. J. Audubon F.R.S. F.L.S. Engraved, Printed & Coloured by R. Havell, London 1833

烟囱雨燕

学　名：*Chaetura pelagica*

英文名：Chimney Swift

1. 雄性　2. 雌性

No. 32. PLATE CLIX.

Cardinal Grosbeak

FRINGILLA CARDINALIS, Bonap

Male 1. Female 2.

Wild Almond

Drawn from Nature by J. J. Audubon F.R.S. F.L.S.

Engraved, Printed & Coloured by R. Havell, London 1833

主红雀

学　名：*Cardinalis cardinalis*

英文名：Northern Cardinal

1. 雄性　2. 雌性

Black-capped Titmouse

PARUS ATRICAPILLUS, Linn.

Male 1. Female 2.

Supple-jack

Drawn from Nature by J. J. Audubon F.R.S. F.L.S.

Engraved, Printed & Coloured by R. Havell, London 1833

卡罗山雀

学　名：*Parus carolinensis*

英文名：Carolina Chickadee

1. 雄性　2. 雌性

No 33. PLATE CLXI.

Brasilian Caracara Eagle

Drawn from Nature by J. J. Audubon F.R.S. F.L.S. POLYBORUS VULGARIS Engraved, Printed & Coloured by R. Havell, London 1833

凤头巨隼

学　名：*Caracara cheriway*

（原为 *Plemeus* 属）

英文名：Common Caracara

Zenaida dove

COLUMBA ZENAIDA

Male 1. Female 2.

Anona

Drawn from Nature by J. J. Audubon F.R.S. F.L.S.

Engraved, Printed & Coloured by R. Havell, London 183.

鸣哀鸽

学　名：*Zenaida aurita*

英文名：Zenaida Dove

1. 雄性　2. 雌性

Palm Warbler

SYLVIA PALMARUM

Male 1. Young 2.

Wild Orange

Drawn from Nature by J. J. Audubon F.R.S. F.L.S.

Engraved, Printed & Coloured by R. Havell, London 1833

棕榈林莺

学　名：*Setophaga palmarum*

（原为 *Dendroica* 属）

英文名：Palm Warbler

1. 雄性　2. 亚成体

Tawny Thrush

TURDUS WILSONII *Male* *Habenaria Lacera – Cornus Canadensis*

Drawn from Nature by J. J. Audubon F.R.S. F.L.S. Engraved, Printed & Coloured by R. Havell, London 1833

棕夜鸫（雄性）

学　名：*Catharus fuscescens*

英文名：Veery

Bachmans Finch

FRINGILLA BACHMANI

Male

Pinckneya Pubens

Drawn from Nature by J. J. Audubon F.R.S. F.L.S.

Engraved, Printed & Coloured by R. Havell, London 1833

巴氏猛雀鹀（雄性）

学 名：*Peucaea aestivalis*

（原为 *Aimophila* 属）

英文名：Bachman's Sparrow

Rough-legged Falcon

FLACO LAGOPUS

Drawn from Nature by J. J. Audubon F.R.S. F.L.S.

Male

Engraved, Printed & Coloured by R. Havell, London 183

毛脚鵟（雄性）

学　名：*Buteo lagopus*

英文名：Rough-legged Hawk

Key-west Dove COLUMBA MONTANA

Male 1. Female 2.

Drawn from Nature by J. J. Audubon F.R.S. F.L.S.

Engraved, Printed & Coloured by R. Havell, London 1833

绿顶鹑鸠

学　名：*Geotrygon chrysia*

英文名：Key West Quail-dove

1. 雄性　2. 雌性

Forked-tailed Flycatcher MUSCICAPA SAVANA *Male* *Gordonia Lasianthus*

Drawn from Nature by J. J. Audubon F.R.S. F.L.S.

Engraved, Printed & Coloured by R. Havell, London 183

叉尾王霸鹟（雄性）

学　名：*Tyrannus savana*

英文名：Fork-tailed Flycatcher

Drawn from Nature by J. J. Audubon F.R.S. F.L.S. *Mangrove Cuckoo* COCCYZUS SENICULUS. *Male* Engraved, Printed & Coloured by R. Havell, 1833

红树美洲鹃（雄性）

学　名：*Coccyzus minor*

英文名：Mangrove Cuckoo

No. 34. PLATE CLXX.

Gray Tyrant

TYRANNUS GRISENS

Drawn from Nature by J. J. Audubon F.R.S. F.L.S. *Agati grandiflora* Engraved, Printed & Coloured by R. Havell, 1833

灰王霸鹟

学　名：*Tyrannus dominicensis*

英文名：Gray Kingbird

Barn Owl

STRIX FLAMMEA

Male 1. Female 2.

Ground Squirrel, Sciurus Shiatus

Drawn from Nature by J. J. Audubon F.R.S. F.L.S.

Engraved, Printed & Coloured by R. Havell, 1833

仓 鸮

学　名：*Tyto alba*

英文名：Western Barn Owl

1. 雄性　2. 雌性

No. 35.

PLATE CLXXII.

Blue-headed Pigeon

COLUMBA CYANOCEPHALA

Male 1. Female 2.

Drawn from Nature by J. J. Audubon F.R.S. F.L.S.

Engraved, Printed & Coloured by R. Havell, 1833

蓝头鹑鸠

学　名：*Starnoenas cyanocephala*

英文名：Blue-headed Quail-dove

1　雄性　2　雌性

Barn Swallow

HIRUNDO AMERICANA

Male 1. Female 2.

Drawn from Nature by J. J. Audubon F.R.S. F.L.S.

Engraved, Printed & Coloured by R. Havell, 1833

家 燕

学 名：*Hirundo rustica*

英文名：Barn Swallow

1. 雄性 2. 雌性

MUSCICAPA INORNATA
Male 1. Female 2.
Pinus Balsamea. Fir Balsam

Drawn from Nature by J. J. Audubon F.R.S. F.L.S.

Engraved, Printed & Coloured by R. Havell, London 1833

绿胁绿霸鹟

学　名：*Contopus cooperi*

英文名：Olive-sided Flycatcher

1. 雄性　2. 雌性

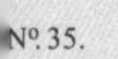

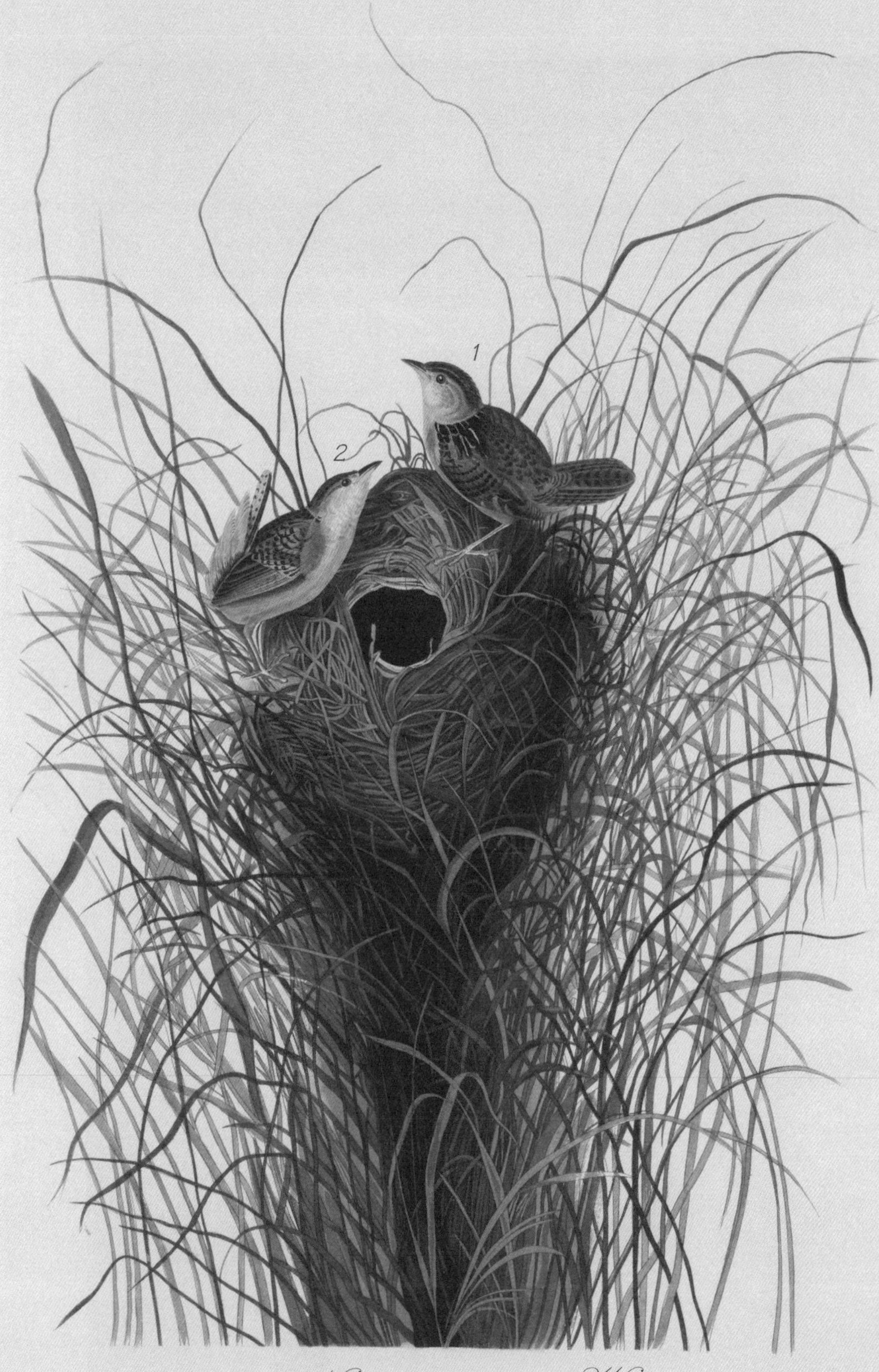

Nuttalls lesser-marsh Wren

TROGLODITES BREVIROSTRIS

Drawn from Nature by J. J. Audubon F.R.S. F.L.S.

Male 1. Female 2.

Engraved, Printed & Coloured by R. Havell, 1833

短嘴沼泽鹪鹩

学　名：*Cistothorus platensis*

英文名：Sedge Wren

1. 雄性　2. 雌性

No. 36.　　PLATE CLXXVI.

Spotted Grous

TETRAO CANADENSIS

Males 1. Females 2.

Trillium pictum, Streptopus distortus

Drawn from Nature by J. J. Audubon F.R.S. F.L.S.

Engraved, Printed & Coloured by R. Havell, London 1833

枞树镰翅鸡

学　名：*Falcipennis canadensis*

（原为 *Dendragapus* 属）

英文名：Spruce Grouse

1. 雄性　2. 雌性

White-crowned Pigeon

COLUMBA LEUCOCEPHALA

Male 1. Female 2.

rawn from Nature by J. J. Audubon F.R.S. F.L.S. *Coidia Sebestena* Engraved, Printed & Coloured by R. Havell, London 1833

白顶鸽

学　名：*Patagioenas leucocephala*

（原为 *Columba* 属）

英文名：White-crowned Pigeon

1. 雄性　2. 雌性

No. 36. PLATE CLXXVII

Orange-crowned Warbler SYLVIA CELATA *Male 1. Female 2. Vaccinium*

Drawn from Nature by J. J. Audubon F.R.S. F.L.S. Engraved, Printed & Coloured by R. Havell

橙冠虫森莺

学　名：*Oreothlypis celata*

（原为 *Helmintophila, Leiothlypis, Vermivora* 属）

英文名：Orange-crowned Warbler

1. 雄性　2. 雌性

Wood Wren

TROGLODYTES AMERICANA

Male

Smilacina borealis

Drawn from Nature by J. J. Audubon F.R.S. F.L.S.　　Engraved, Printed & Coloured by R. Havell, 1833

莺鹪鹩（雄性）

学　名：*Troglodytes aedon*

英文名：House Wern

Pine Finch

FRINGILLA PINUS

Male 1. Female 2.

Pinus pendula. Black Larch

Drawn from Nature by J. J. Audubon F.R.S. F.L.S　　Engraved, Printed & Coloured by R. Havell, 183

松金翅雀

学　名：*Spinus pinus*

（原为 *Carduelis* 属）

英文名：Pine Siskin

1. 雄性　2. 雌性

Golden Eagle AQUILA CHRYSAETOS *Female adult* *Northern Hare*

Drawn from Nature by J. J. Audubon F.R.S. F.L.S. Engraved, Printed & Coloured by R. Havell, 1833

金　雕（成年雌性）

学　名：*Aquila chrysaetos*

英文名：Golden Eagle

N° 37. PLATE CLXXXII.

Ground Dove COLUMBA PASSERINA, Linn. *Males 1, 2, 3. Female 4. Young 5. Wild Orange*

Drawn from Nature by J. J. Audubon F.R.S. F.L.S. Engraved, Printed & Coloured by R. Havell, 1833

地　鸠

学　名：*Columbina passerina*

英文名：Common Ground Dove

1,2,3.　雄性　4.　雌性　5.　亚成体

Golden crested-Wren REGULUS CRISTATUS, Vieill *Male 1. Female 2. Thalia dealbata*

Drawn from Nature by J. J. Audubon F.R.S. F.L.S. Engraved, Printed & Coloured by R. Havell

金冠戴菊

学　名：*Regulus satrapa*

英文名：Golden-crowned Kinglet

1. 雄性　2. 雌性

No 37. PLATE CLXXXIV.

Mangrove Humming Bird

TROCHILUS MANGO

Males 1, 2, 3. Females 4, 5.

Tecoma grandiflora

Drawn from Nature by J. J. Audubon F.R.S. F.L.S. Engraved, Printed & Coloured by R. Havell, 1833

黑喉芒果蜂鸟

学　名：*Anthracothorax nigricollis*

英文名：Black-throated Mango

1,2,3. 雄性　4,5. 雌性

No. 37. PLATE CLXXXV.

Bachman's Warbler SYLVIA BACHMANII, Aud. *Male 1. Female 2. Gordonia pubescens*

Drawn from Nature by J. J. Audubon F.R.S. F.L.S. Engraved, Printed & Coloured by R. Havell, 1833

黑胸虫森莺

学　名：*Vermivora bachmanii*

英文名：Bachman's Warbler

1. 雄性　2. 雌性

No. 38.

PLATE CLXXXVI.

1

3

2

Drawn from Nature by J. J. Audubon F.R.S. L.S.

Pinnated Grous TETRAO CUPIDO, Linn. *Males 1, 2. Female 3. Lilium Superbum*

Engraved, Printed & Coloured by R. Havell, 1834

草原松鸡

学　名：*Tympanuchus cupido*

英文名：Greater Prairie Chicken

1.2. 雄性　3. 雌性

Boat-tailed Grackle

QUISCALUS MAJOR, Vieill

Male 1. Female 2.

Drawn from Nature by J. J. Audubon F.R.S. L.S.

Live Oak—Quercus virens

Engraved, Printed & Coloured by R. Havell, 1834

宽尾拟八哥

学　名：*Quiscalus major*

英文名：Boat-tailed Grackle

1. 雄性　2. 雌性

Nº 38.

PLATE CLXXXVIII.

Tree Sparrow

FRINGILLA CANADENSIS, Lath.

Drawn from Nature by J. J. Audubon F.R.S. L.S.

Berberis Canadensis

Engraved, Printed & Coloured by R. Havell, 1834

美洲树雀鹀

学　名：*Spizelloides arborea*

（原为 *Spizella* 属）

英文名：American Tree Sparrow

Snow Bunting

EMBERIZA NIVALIS, Linn.

Drawn from Nature by J. J. Audubon F.R.S. F.L.S.　　*Adult 1, 2. Young 3.*　　Engraved, Printed & Coloured by R. Havell, 1834

雪 鹀

学　名：*Plectrophenax nivalis*

英文名：Snow Bunting

1,2. 成体　3. 亚成体

黄腹吸汁啄木鸟

学　名：*Sphyrapicus varius*

英文名：Yellow-bellied Sapsucker

1. 雄性　2. 雌性

Willow Grous or Large Ptarmigan

TETRAO SALICETI, Temm.

Male 1. Female 2. & Young

Labrdor Tea, Sweet pea

Drawn from Nature by J. J. Audubon F.R. S. F.L. S.

Engraved, Printed & Coloured by R. Havell, 1834

柳雷鸟

学　名：*Lagopus lagopus*

英文名：Willow Ptarmigan

1．雄性　2．雌性和雏鸟

Great American Shrike or Butcher Bird LANIUS SEPTENTUONALIS *Male 1. F. 2. Summer Plumage Do. 3. Young or Winter Do.*
Crataqus apiifolia

Drawn from Nature by J. J. Audubon F.R.S. L.S.

Engraved, Printed & Coloured by R. Havell, 183

灰伯劳

学　名：*Lanius excubitor*

英文名：Great Grey Shrike

1. 雄性　2. 雌性　3. 夏季的雄性　4. 冬季的雌性或亚成体

Lincoln Finch

FRINGILLA LINCOLNII

Male 1. Female 2.

Drawn from Nature by J. J. Audubon F.R.S. F.L.S.　*Cornus Suifsica, Rubus Chamarus, Kalmia glauca*　Engraved, Printed & Coloured by R. Havell, 1834

林氏带鹀

学　名：*Melospiza lincolnii*

英文名：Lincoln's Sparrow

1. 雄性　2. 雌性

Canadian Titmouse

PARUS HUDSONICUS

Drawn from Nature by J. J. Audubon F.R.S. F.L.S. *Male 1. Female 2. Young 3.* Engraved, Printed & Coloured by R. Havell, 183

北山雀

学　名：*Poecile hudsonicus*

（原为 *Parus* 属）

英文名：Boreal Chickadee

1．雄性　2．雌性　3．亚成体

No 39. PLATE CXCV.

Ruby crowned Wren

REGULUS CALENDULA, Stephens

Male 1. Female 2. Summer plumage

Kalmia Angustifolia

Drawn from Nature by J. J. Audubon F.R.S. F.L.S.

Engraved, Printed & Coloured by R. Havell, 1834

红冠戴菊

学　名：*Regulus calendula*

英文名：Ruby-crowned Kinglet

1. 夏季的雄性　2. 夏季的雌性

Labrador Falcon

FALCO LABRADORA

Drawn from Nature by J. J. Audubon F.R.S. F.L.S.

Male 1. Female 2. adult

Engraved, Printed & Coloured by R. Havell, 18

矛　隼

学　名：*Falco rusticolus*

英文名：Gyrfalcon

1．成年雄性　2．成年雌性

American Crossbill LOXIA CURVIROSTRA, Linn.
Male adult 1. Young Male 2, 3. Female adult 4. Young Female 5.
Hemlock

Drawn from Nature by J. J. Audubon F.R.S. L.S. Engraved, Printed & Coloured by R. Havell, 1834

红交嘴雀

学　名：*Loxia curvirostra*

英文名：Red Crossbill

1．成年雄性　2,3．雄性，亚成体　4．成年雌性　5．雌性，亚成体

No. 40. PLATE CXCVIII.

Brown headed Worm eating Warbler

SYLVIA SWAINSONII

Drawn from Nature by J. J. Audubon F.R.S. F.L.S. *Azalia Calendula — Orange coloured Azalia* Engraved, Printed & Coloured by R. Havel

白眉食虫莺

学　名：*Limnothlypis swainsonii*

英文名：Swainson's Warbler

Little Owl

STRIX ACADICA, Gm.

Male 1. Female 2.

Common Mouse

Drawn from Nature by J. J. Audubon F.R.S. F.L.S.

Engraved, Printed & Coloured by R. Havell, 1834

棕榈鬼鸮

学　名：*Aegolius acadicus*

英文名：Northern Saw-whet Owl

1. 雄性　2. 雌性

No. 40.

PLATE CC.

Shore Lark

ALAUDA ALPESTRIS, Z.

Drawn from Nature by J. J. Audubon F.R.S. F.L.S.

Male adult Summer plumage 1. F. 2. Male Winter plumage 3. Young 4, 5, 6.

Engraved, Printed & Coloured by R. Havell

角百灵

学　名：*Eremophila alpestris*

英文名：Horned Lark

1　夏季的成年雄性　2　雌性　3　冬季的雄性　4 5 6　均为亚成体

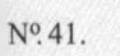
No 41.

PLATE CCI.

Canada Goose

ANSER CANADENSIS, Vieill

Drawn from Nature by J. J. Audubon F.R.S. F.L.S.

Male 1. Female 2.

Engraved, Printed & Coloured by R. Havell

加拿大黑雁

学　名：*Branta canadensis*

英文名：Canada Goose

1. 雄性　2. 雌性

N°.41. PLATE CCII.

Red-throated Diver COLUMBUS SEPTENTRIONALIS *Male adult summer plumage 1. Do. Winter plumage 2. Adult Female 3. Young 4.*

Drawn from Nature by J. J. Audubon F.R.S. F.L.S.

Engraved, Printed & Coloured by R. Havell, 1834

红喉潜鸟

学　名：*Gavia stellata*

英文名：Red-throated Diver

1　夏季的成年雄性　2　冬季的雄性　3　成年雌性　4　雏鸟

Drawn from Nature by J. J. Audubon F.R.S. F.L.S. *Fresh Water Marsh Hen* RALLUS ELEGANS, Aud. *Male spring plumage 1. Young autumnal plumage 2.* Engraved, Printed & Coloured by R. Havell, 1834

王秧鸡

学　名：*Rallus elegans*

英文名：King Rail

1. 春季的雄性　2. 秋季的亚成体

Salt Water Marsh Hen RALLUS CREPITANS, Gm. *1. Male adult spring plumage 2. Female*

Drawn from Nature by J. J. Audubon F.R.S. F.L.S. Engraved, Printed & Coloured by R. Havell, 1834

长嘴秧鸡

学　名：*Rallus longirostris*

英文名：Mangrove Rail

1. 春季的成年雄性　2. 雌性

Drawn from Nature by J. J. Audubon F.R.S. F.L.S. *Virginia Rail* RALLUS VIRGINIANUS, L. *Male 1. Female 2. Young autumnal plumage 3.* Engraved, Printed & Coloured by R. Havell, 1834

弗吉尼亚秧鸡

学　名：*Rallus limicola*

英文名：Virginia Rail

1. 雄性　2. 雌性　3. 秋季的亚成体

ANAS SPONSA, L.

1, 2. Males 3, 4. Females

Drawn from Nature by J. J. Audubon F.R.S. F.L.S. *Platanus occidentalis. - Button Wood Tree* Engraved, Printed & Coloured by R. Havell, 1834

林鸳鸯

学　名：*Aix sponsa*

英文名：Wood Duck

1,2.　雄性　3,4.　雌性

Booby Gannet

Drawn from Nature by J. J. Audubon F.R.S. F.L.S.

SULA FUSCA

Engraved, Printed & Coloured by R. Havell, 1834

褐鲣鸟

学　名：*Sula leucogaster*

英文名：Brown Booby

Nº 42.

PLATE CCVIII.

Drawn from Nature by J. J. Audubon F.R.S. F.L.S.

Esquimaux Curlew NUMENIUS BOREALIS, Lath. *Male 1. Female 2.*

Engraved, Printed & Coloured by R. Havell, 1834

极北杓鹬

学　名：*Numenius borealis*

英文名：Eskimo Curlew

1　雄性　2　雌性

Drawn from Nature by J. J. Audubon F.R.S. F.L.S.

Wilson's Plover CHARADRIUS WILSONIUS *1. Male 2. Female*

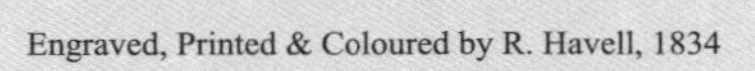
Engraved, Printed & Coloured by R. Havell, 1834

厚嘴鸻

学　名：*Charadrius wilsonia*

英文名：Wilson’s Plover

1. 雄性　2. 雌性

Nº 42. PLATE CCX.

Drawn from Nature by J. J. Audubon F.R.S. F.L.S. *Least Bittern* ARDEA EXILIS, Gm. *1. Male 2. Female 3. Young* Engraved, Printed & Coloured by R. Havell, 1834

姬苇鳽

学　名：*Ixobrychus exilis*

英文名：Least Bittern

1　雄性　2　雌性　3　亚成体

rawn from Nature by J. J. Audubon F.R.S. F.L.S. *Great blue Heron* ARDEA HERODIAS *Male* Engraved, Printed & Coloured by R. Havell, 1834

大蓝鹭（雄性）

学　名：*Ardea herodias*

英文名：Great Blue Heron

Drawn from Nature by J. J. Audubon F.R.S. F.L.S. *Common Gull* LARUS CANUS *1. Adult 2. Young* Engraved, Printed & Coloured by R. Havell, 1834

环嘴鸥

学　名：*Larus delawarensis*

英文名：Ring-billed Gull

1. 成体 2. 亚成体

Drawn from Nature by J. J. Audubon F.R.S. F.L.S.

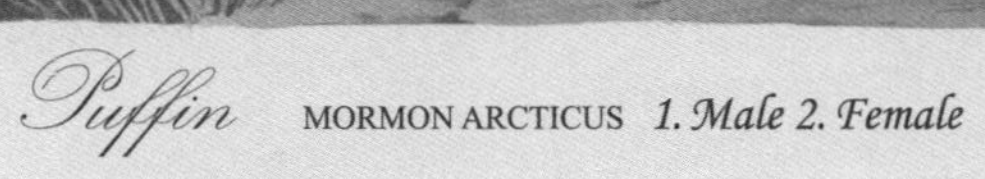

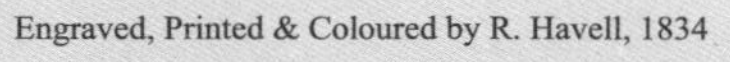

北极海鹦

学　名：*Fratercula arctica*

英文名：Atlantic Puffin

1．雄性　2．雌性

Nº 43. PLATE CCXIV.

Drawn from Nature by J. J. Audubon F.R.S. F.L.S.

Razor Bill ALCA TORDA *1. Male 2. Female*

Engraved, Printed & Coloured by R. Havell, 1834

刀嘴海雀

学　名：*Alca torda*

英文名：Razorbill

1. 雄性　2. 雌性

Hyperborean phalarope PHALAROPUS HYPERBOREUS, Lath. *1. Male adult spring plumage 2. Female Do. 3. Young autumnal plumage*

Drawn from Nature by J. J. Audubon F.R.S. F.L.S.

Engraved, Printed & Coloured by R. Havell, 1834

红颈瓣蹼鹬

学　名：*Phalaropus lobatus*

英文名：Red-necked Phalarope

1. 春季的成年雄性　2. 雌性　3. 秋季的亚成体

No. 44.

PLATE CCXVI.

Drawn from Nature by J. J. Audubon F.R.S. F.L.S. *Wood Ibiss* TANTALUS LOCULATOR Engraved, Printed & Coloured by R. Havell, 1834

黑头鹮鹳

学　名：*Mycteria americana*

英文名：Wood Stork

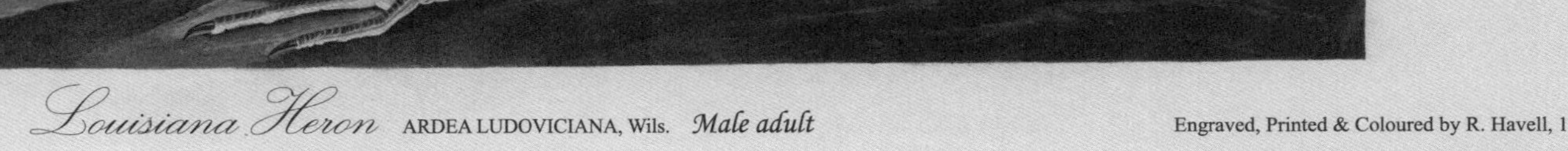

Drawn from Nature by J. J. Audubon F.R.S. F.L.S.

Louisiana Heron ARDEA LUDOVICIANA, Wils. *Male adult*

Engraved, Printed & Coloured by R. Havell, 1834

三色鹭（成年雄性）

学　名：*Egretta tricolor*

英文名：Tricolored Heron

No. 44. PLATE CCXVIII.

Drawn from Nature by J. J. Audubon F.R.S. F.L.S.

Foolish Guillemot URIA TROILE, Lath. *1. Adult summer plumage Male 2. Female*

Engraved, Printed & Coloured by R. Havell, 1834

崖海鸦

学　名：*Uria aalge*

英文名：Common Murre

1　夏季的成年雄性　2　雌性

Black Guillemot

URIA GRYLLE, Lath.

Drawn from Nature by J. J. Audubon F.R.S. F.L.S.

1. Adult summer plumage 2. Do. Winter plumage 3. Young

Engraved, Printed & Coloured by R. Havell, 1834

白翅斑海鸽

学　名：*Cepphus grylle*

英文名：Black Guillemot

1. 夏季的成体　2. 冬季的成体　3. 雏鸟

No. 44.

PLATE CCXX.

Drawn from Nature by J. J. Audubon F.R.S. F.L.S.

Piping Plover CHARADRIUS MELODUS *1. Male 2. Female*

Engraved, Printed & Coloured by R. Havell, 1834

笛 鸻

学　名：*Charadrius melodus*

英文名：Piping Plover

1. 雄性　2. 雌性

Drawn from Nature by J. J. Audubon F.R.S. F.L.S.

Mallard Duck ANAS BOSCHAS, L. *1. Males 2. Females*

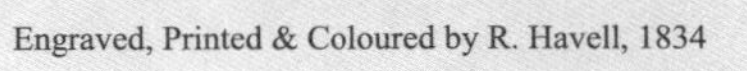

Engraved, Printed & Coloured by R. Havell, 1834

绿头鸭

学　名：*Anas platyrhynchos*

英文名：Mallard

1. 雄性　2. 雌性

Nº 45.

PLATE CCXXII.

Drawn from Nature by J. J. Audubon F.R.S. F.L.S.

White Ibis IBIS ALBA *1. Adult 2. Young in Autumn*

Engraved, Printed & Coloured by R. Havell, 1834

美洲白鹮

学　名：*Eudocimus albus*

英文名：American White Ibis

1. 成体 2. 秋季的亚成体

Pied oyster-catcher

Drawn from Nature by J. J. Audubon F.R.S. F.L.S.

HÆMATOPUS OSTRALEGUS, L.

Engraved, Printed & Coloured by R. Havell, 1834

美洲蛎鹬

学　名：*Haematopus palliatus*

英文名：American Oystercatcher

No.45.

PLATE CCXXIV.

2

1

Drawn from Nature by J. J. Audubon F.R.S. F.L.S.

Kittiwake Gull LARUS TRIDACTYLUS, L. *1. Adult 2. Young*

Engraved, Printed & Coloured by R. Havell, 1834

三趾鸥

学　名：*Rissa tridactyla*

英文名：Black-legged Kittiwake

1　成体　2　亚成体

1

2

Drawn from Nature by J. J. Audubon F.R.S. F.L.S.

Kildeer Plover CHARADRIUS VOCIFERUS *1. Male 2. Female*

Engraved, Printed & Coloured by R. Havell, 1834

双领鸻

学　名：*Charadrius vociferus*

英文名：Killdeer

1. 雄性　2. 雌性

Drawn from Nature by J. J. Audubon F.R.S. F.L.S. *Hooping Crane* GRUS AMERICANA *Adult Male* Engraved, Printed & Coloured by R. Havell, 183

美洲鹤（成年雄性）

学　名：*Grus americana*

英文名：Whooping Crane

Drawn from Nature by J. J. Audubon F.R.S. F.L.S. *Pin tailed Duck* ANAS ACUTA *Male 1. Female 2.* Engraved, Printed & Coloured by R. Havell, 1834

针尾鸭

学　名：*Anas acuta*

英文名：Northern Pintail

1. 雄性　2. 雌性

Drawn from Nature by J. J. Audubon F.R.S. F.L.S. *American Green winged Teal* ANAS CAROLINENSIS, Lath. *1. Male 2. Female* Engraved, Printed & Coloured by R. Havell, 1834

绿翅鸭

学　名：*Anas crecca*

英文名：Eurasian Teal

1　雄性　2　雌性

Drawn from Nature by J. J. Audubon F.R.S. F.L.S.　*Scaup Duck* FULIGULA MARILA *1. Male 2. Female*　Engraved, Printed & Coloured by R. Havell, 1834

小潜鸭

学　名：*Aythya affinis*

英文名：Lesser Scaup

1. 雄性　2. 雌性

Drawn from Nature by J. J. Audubon F.R.S. F.L.S. *Ruddy Plover* TRINGA ARENARIA *1. Male 2. Female* Engraved, Printed & Coloured by R. Havell, 1834

三趾滨鹬

学　名：*Calidris alba*

英文名：Sanderling

1. 雄性　2. 雌性

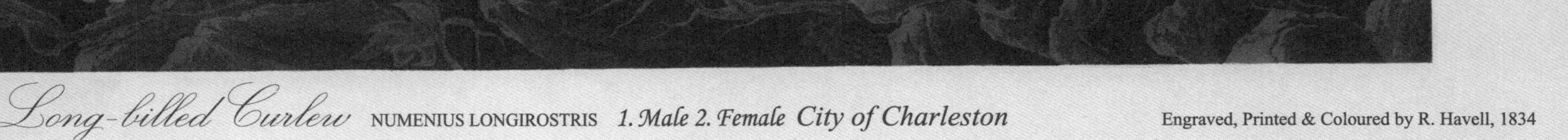

Drawn from Nature by J. J. Audubon F.R.S. F.L.S. *Long-billed Curlew* NUMENIUS LONGIROSTRIS *1. Male 2. Female City of Charleston* Engraved, Printed & Coloured by R. Havell, 1834

长嘴杓鹬

学　名：*Numenius americanus*

英文名：Long-billed Curlew

1. 雄性　2. 雌性

No.47. PLATE CCXXXII.

Drawn from Nature by J. J. Audubon F.R.S. F.L.S. *Hooded Merganser* MERGUS CUCULLATUS *1. Male 2. Female* Engraved, Printed & Coloured by R. Havell, 1834

棕胁秋沙鸭

学　名：*Lophodytes cucullatus*

英文名：Hooded Merganser

1. 雄性　2. 雌性

Drawn from Nature by J. J. Audubon F.R.S. F.L.S.

Sora or Rail RALLUS CAROLINUS, L. *1. Male 2. Female 3. Young*

Engraved, Printed & Coloured by R. Havell, 1834

黑脸田鸡

学　名：*Porzana carolina*

英文名：Sora

1. 雄性　2. 雌性　3. 亚成体

Nº 47.

PLATE CCXXXIV.

Drawn from Nature by J. J. Audubon F.R.S. F.L.S.

Tufted Duck FULIGULA RUFITORQUES, Bonap *1. Male 2. Female*

Engraved, Printed & Coloured by R. Havell, 1834

环颈潜鸭

学　名：*Aythya collaris*

英文名：Ring-necked Duck

1. 雄性 2. 雌性

乌燕鸥

学　名：*Onychoprion fuscata*

（原为 *Sterna* 属）

英文名：Sooty Tern

Nº 48.

PLATE CCXXXVI.

Drawn from Nature by J. J. Audubon F.R.S. F.L.S.

Night Heron or Qua bird ARDEA NYCTICORAX, L. *Adult 1. and Young 2.*

Engraved, Printed & Coloured by R. Havell, 1835

夜　鹭

学　名：*Nycticorax nycticorax*

英文名：Black-crowned Night Heron

1　成体　2　亚成体

Great Esquimaux Curlew

Drawn from Nature by J. J. Audubon F.R.S. F.L.S.

NUMENIUS HUDSONICUS, Lath.

Engraved, Printed & Coloured by R. Havell, 1835

中杓鹬

学　名：*Numenius phaeopus*

英文名：Whimbrel

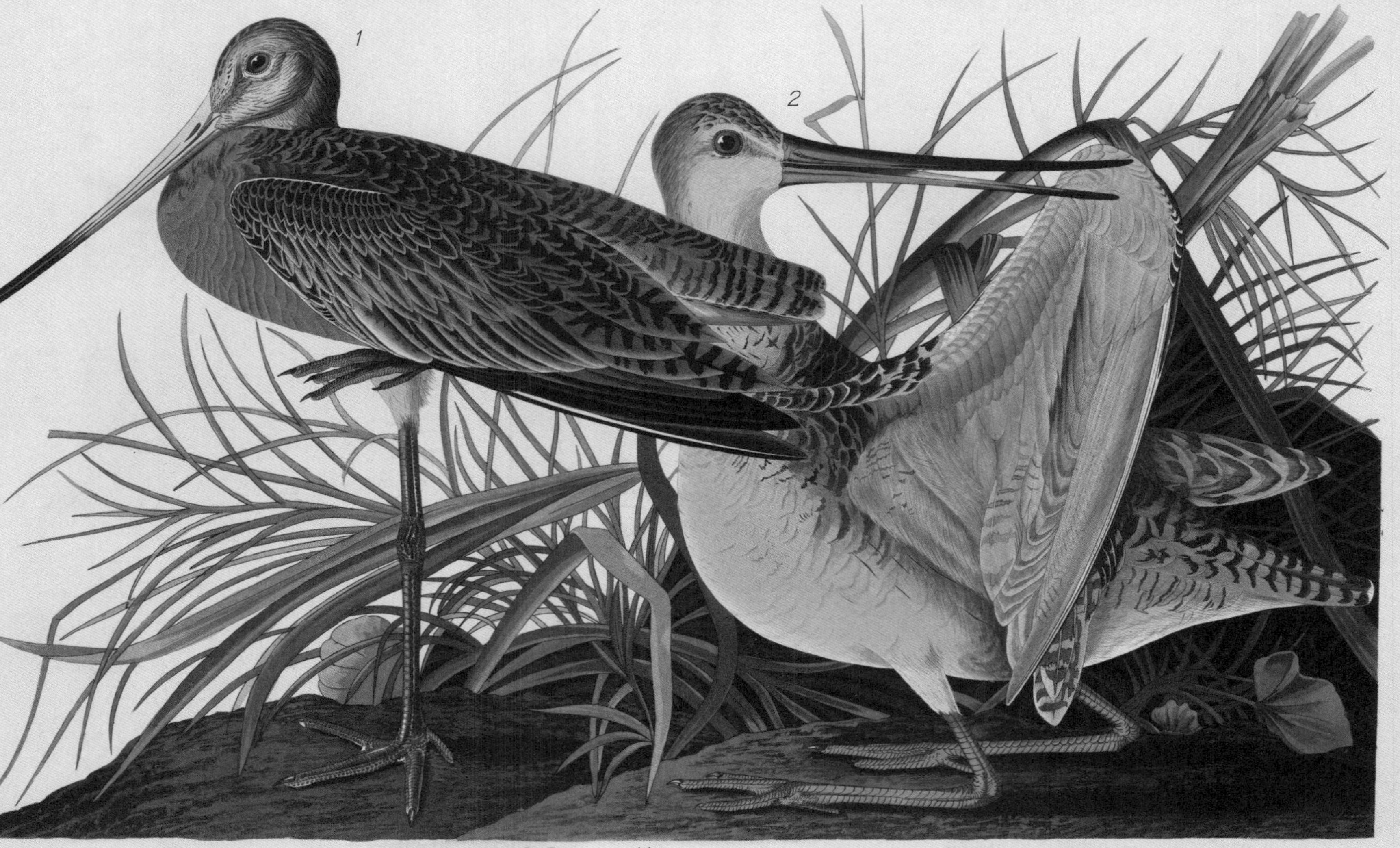

Great Marbled Godwit

LIMOSA FEDOA, Vieill

1. Male 2. Female

Drawn from Nature by J. J. Audubon F.R.S. F.L.S.

Engraved, Printed & Coloured by R. Havell, London 1835

云斑塍鹬

学　名：*Limosa fedoa*

英文名：Marbled Godwit

American Coot

Drawn from Nature by J. J. Audubon F.R.S. F.L.S.

FULICA AMERICANA, Gm.

Engraved, Printed & Coloured by R. Havell, London 1835

美洲骨顶

学　名：*Fulica americana*

英文名：American Coot

Roseate Tern

Drawn from Nature by J. J. Audubon F.R.S. F.L.S. STERNA DOUGALLII, Mont. Engraved, Printed & Coloured by R. Havell, London 18

粉红燕鸥

学　名：*Sterna dougallii*

英文名：Roseate Tern

№.49. PLATE CCXLI.

Black Backed Gull

Drawn from Nature by J. J. Audubon F.R.S. F.L.S. LARUS MARINUS Engraved, Printed & Coloured by R. Havell, 1835

大黑背鸥

学　名：*Larus marinus*

英文名：Great Black-backed Gull

Snowy Heron or White Egret ARDEA CANDIDISSIMA, Gm.

Male adult Spring plumage

Drawn from Nature by J. J. Audubon F.R.S. F.L.S. *Rice plantation. South Carolina* Engraved, Printed & Coloured by R. Havell, 183

雪　鹭（春季的成年雄性）

学　名：*Egretta thula*

英文名：Snowy Egret

American Snipe Male 1. Female 2,3.

SCOLOPAX WILSONII

Drawn from Nature by J. J. Audubon F.R.S. F.L.S.

South Carolina Plantation near Charleston

Engraved, Printed & Coloured by R. Havell, London 1835

扇尾沙锥

学　名：*Gallinago gallinago*

英文名：Common Snipe

1．雄性　2,3．雌性

黑水鸡（成年雄性）

学　名：*Gallinula galeata*

（原为 *Chloropus* 属）

英文名：Common Gallinule

Drawn from Nature by J. J. Audubon F.R.S. F.L.S. URIA BRUNNICHII Engraved, Printed & Coloured by R. Havell, London 1835

厚嘴崖海鸦

学　名：*Uria lomvia*

英文名：Thick-billed Murre

Drawn from Nature by J. J. Audubon F.R.S. F.L.S. *Eider Duck* *Male 1. Female 2.* FULIGULA MOLLISSIMA Engraved, Printed & Coloured by R. Havell, London 1835

欧绒鸭

学　名：*Somateria mollissima*

英文名：Common Eider

1　雄性　2　雌性

Velvet Duck

Drawn from Nature by J. J. Audubon F.R.S. F.L.S.

FULIGULA FUSCA *Male 1. Female 2.*

Engraved, Printed & Coloured by R. Havell, London 1835

斑脸海番鸭

学　名：*Melanitta fusca*

英文名：Velvet Scoter

1. 雄性　2. 雌性

American Pied-bill Dobchick

Drawn from Nature by J. J. Audubon F.R.S. F.L.S.

PODICEPS CAROLINENSIS

Engraved, Printed & Coloured by R. Havell, London 1835

斑嘴巨䴙䴘

学　名：*Podilymbus podiceps*

英文名：Pied-billed Grebe

Tufted Auk

Drawn from Nature by J. J. Audubon F.R.S. F.L.S.

MORMON CIRRHATUS, Lath. *Male 1. Female 2.*

Engraved, Printed & Coloured by R. Havell, London 1835

簇羽海鹦

学　名：*Fratercula cirrhata*

英文名：Tufted Puffin

1. 雄性 2. 雌性

Arctic Tern

Drawn from Nature by J. J. Audubon F.R.S. F.L.S. STERNA ARCTICA Engraved, Printed & Coloured by R. Havell, London 1835

北极燕鸥

学　名：*Sterna paradisaea*

英文名：Arctic Tern

褐鹈鹕（成年雄性）

学　名：*Pelecanus occidentalis*

英文名：Brown Pelican

No. 51.

PLATE CCLII.

Florida Cormorant

CARBO FLORIDANUS

Male Adult Spring Drefs

View. Florida Keys.

Drawn from Nature by J. J. Audubon F.R.S. F.L.S.

Engraved, Printed & Coloured by R. Havell, 1835

角鸬鹚（成年雄性）

学　名：*Phalacrocorax auritus*

英文名：Double-crested Cormorant

Jager

Drawn from Nature by J. J. Audubon F.R.S. F.L.S.

LESTRIS POMARINA, Temm.

Engraved, Printed & Coloured by R. Havell, 1835

中贼鸥

学　名：*Stercorarius pomarinus*

英文名：Pomarine Skua

Wilson's Phalarope

Drawn from Nature by J. J. Audubon F.R.S. F.L.S.

PHALAROPUS WILSONII, Sabine *Male Adult 1. Female 2.*

Engraved, Printed & Coloured by R. Havell, 1835

细嘴瓣蹼鹬

学　名：*Steganopus tricolor*

英文名：Wilson's Phalarope

Red Phalarope

Drawn from Nature by J. J. Audubon F.R.S. F.L.S.

PHALAROPUS PLATYRHYNCHUS, Temm. *Adult Male 1. Adult Female 2. Winter Plumage 3.*

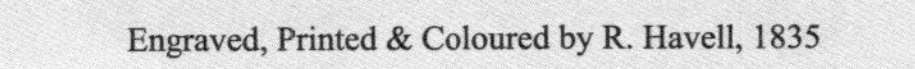

Engraved, Printed & Coloured by R. Havell, 1835

灰瓣蹼鹬

学　名：*Phalaropus fulicarius*

英文名：Red Phalarope

1. 成年雄性　2. 成年雌性　3. 冬季的灰瓣蹼鹬

Nº 52.

PLATE CCLVI.

Purple Heron

ARDEA RUFESCENS, Buff

Drawn from Nature by J. J. Audubon F.R.S. F.L.S.

Adult full spring plumage 1. Young two Years old in spring plumage 2.

Engraved, Printed & Coloured by R. Havell, 1835

棕颈鹭

学　名：*Egretta rufescens*

英文名：Reddish Egret

No. 52. PLATE CCLVII.

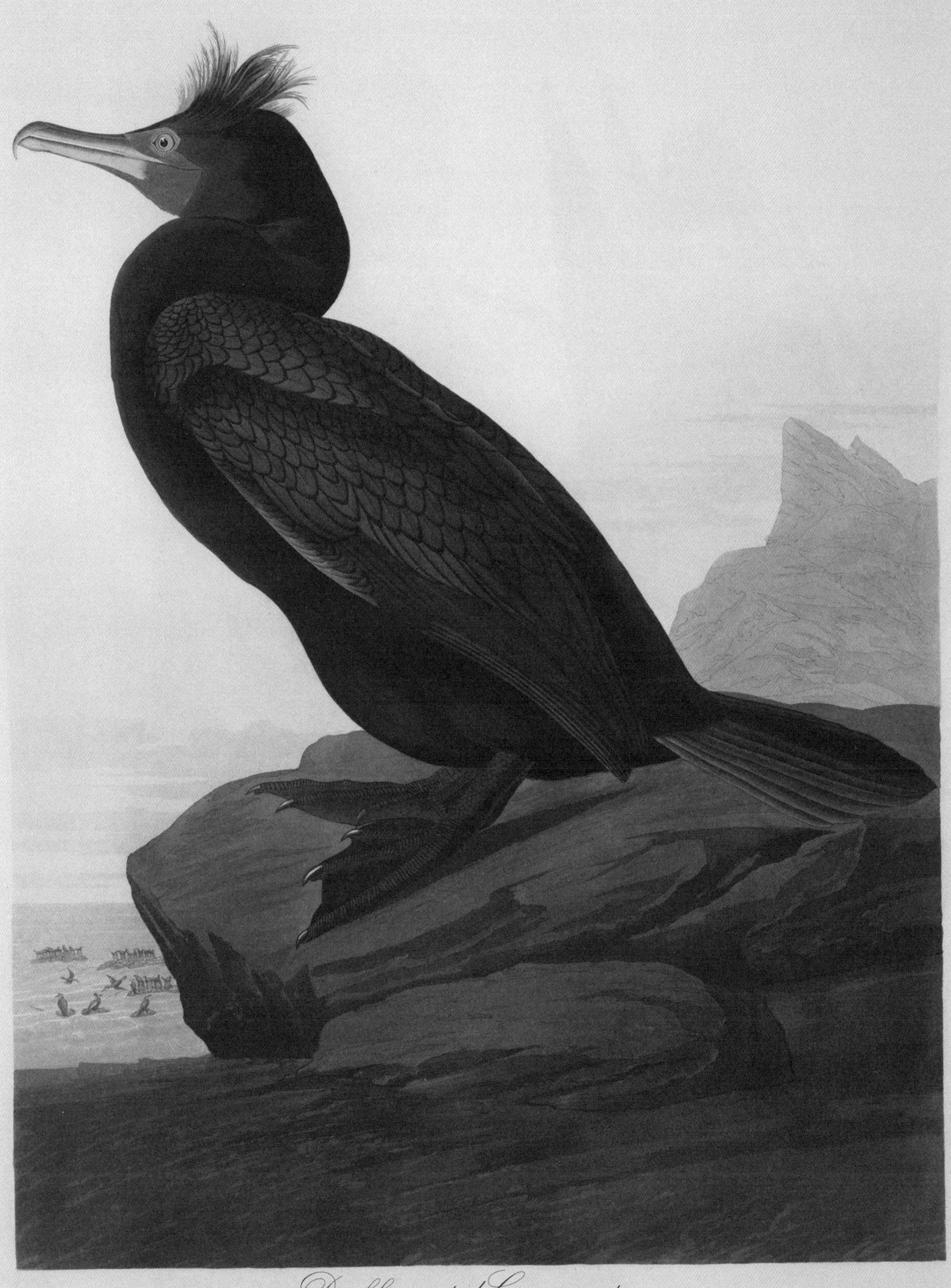

Double-crested Cormorant

PHALACROCORAX DILOPHUS, Swain & Richards

Drawn from Nature by J. J. Audubon F.R.S. F.L.S. *Male adult spring plumage* Engraved, Printed & Coloured by R. Havell, 1835

角鸬鹚（春季的成年雄性）

学　名：*Phalacrocorax auritus*

英文名：Double-crested Cormorant

Hudsonian Godwit

Drawn from Nature by J. J. Audubon F.R.S. F.L.S.   LIMOSA HUDSONICA, Swain & Richards *Male 1. Young Female 2. Summer plumage*   Engraved, Printed & Coloured by R. Havell, 1835

棕塍鹬

学　名：*Limosa haemastica*

英文名：Hudsonian Godwit

1　夏季的雄性　2　夏季的雌性亚成体

Horned Grebe

PODICEPS CORNUTUS, Lath.

Adult Male 1. Female Winter plumage 2.

Drawn from Nature by J. J. Audubon F.R.S. F.L.S.

Engraved, Printed & Coloured by R. Havell, 1835

角䴙䴘

学　名：*Podiceps auritus*

英文名：Horned Grebe

1. 成年雄性　2. 冬季的雌性

No. 52.

PLATE CCLX.

Fork-tail Petrel

Drawn from Nature by J. J. Audubon F.R.S. F.L.S.

THALASSIDROMA LEACHII *Male 1. Female 2.*

Engraved, Printed & Coloured by R. Havell, London 1835

白腰叉尾海燕

学　名：*Oceanodroma leucorhoa*

英文名：Leach's Storm Petrel

1　雄性　2　雌性

Hooping Crane

GRUS AMERICANA

Young

View in the Interior of the Floridas with sand Hills in the distance

Drawn from Nature by J. J. Audubon F.R.S. F.L.S.

Engraved, Printed & Coloured by R. Havell, 1835

沙丘鹤（亚成体）

学　名：*Grus canadensis*

英文名：Sandhill Crane

Nº. 53.

PLATE CCLXII.

1

2

Tropic Bird

PHAETON ÆTHEREUS, Linn.

Male 1. Female 2.

白尾鹲

学　名：*Phaethon lepturus*

英文名：White-tailed Tropicbird

1. 雄性　2. 雌性

Drawn from Nature by J. J. Audubon F.R.S. F.L.S.

Engraved, Printed & Coloured by R. Havell, 1835

Pigmy Curlew

TRINGA SUBARQUATA, Temm.

Adult Male 1. Young 2.

Drawn from Nature by J. J. Audubon F.R.S. F.L.S.

Engraved, Printed & Coloured by R. Havell, 1835

弯嘴滨鹬

学　名：*Calidris ferruginea*

英文名：Curlew Sandpiper

1. 成年雄性　2. 亚成体

No. 53.

PLATE CCLXIV.

Fulmar Petrel

PROCELLARIA GLACIALIS, L.

Male adult Summer plumage

Drawn from Nature by J. J. Audubon F.R.S. F.L.S.

Engraved, Printed & Coloured by R. Havell, 1835

暴风鹱（夏季的成年雄性）

学　名：*Fulmarus glacialis*

英文名：Northern Fulmar

Buff breasted Sandpiper

TRINGA RUFESCENS, Vieill

1. Male 2. Female

Drawn from Nature by J. J. Audubon F.R.S. F.L.S.

Engraved, Printed & Coloured by R. Havell, 1835

黄胸鹬

学　名：*Tryngites subruficollis*

英文名：Buff-breasted Sandpiper

1. 雄性　2. 雌性

No. 54. PLATE CCLXVI.

Common Cormorant

PHALACROCORAX CARBO, Dumont

Male adult spring plumage 1. Female 2. Young 3.

普通鸬鹚

学　名：*Phalacrocorax carbo*

英文名：Great Cormorant

1. 春季的成年雄性　2. 雌性　3. 雏鸟

Drawn from Nature by J. J. Audubon F.R.S. F.L.S.

Engraved, Printed & Coloured by R. Havell. 1835

Arctic Yager

Drawn from Nature by J. J. Audubon F.R.S. F.L.S.　　LESTRIS PARASITICA　　Engraved, Printed & Coloured by R. Havell, 1835

长尾贼鸥

学　名：*Stercorarius longicaudus*

英文名：Long-tailed Jaeger

American Woodcock

Drawn from Nature by J. J. Audubon F.R.S. F.L.S.

SCOLOPAX MINOR, Gmel. *Male 1. Female 2. Young Autumn 3.*

Engraved, Printed & Coloured by R. Havell, London 1835

小丘鹬

学　名：*Scolopax minor*

英文名：American Woodcock

1　雄性　2　雌性　3　秋季的亚成体

Greenshank

Drawn from Nature by J. J. Audubon F.R.S. F.L.S. TOTANUS GLOTTIS, Temm. *View of St Augustine & Spanish Fort East Florida* Engraved, Printed & Coloured by R. Havell, 1835

青脚鹬

学　名：*Tringa nebularia*

英文名：Common Greenshank

No. 54.

PLATE CCLXX.

Stormy Petrel

Drawn from Nature by J. J. Audubon F.R.S. F.L.S.

THALASSIDROMA WILSONII *Male 1. Female 2.*

Engraved, Printed & Coloured by R. Havell, London 1835

黄蹼洋海燕

学　名：*Oceanites oceanicus*

英文名：Wilson's Storm Petrel

1　雄性　2　雌性

华丽军舰鸟（成年雄性）

学　名：*Fregata magnificens*

英文名：Magnificent Frigatebird

Drawn from Nature by J. J. Audubon F.R.S. F.L.S.　*Richardson's Jager*　LESTRIS RICHARDSONII　*Male adult 1. Young in Sepr. 2.*　Engraved, Printed & Coloured by R. Havell, 1835

短尾贼鸥

学　名：*Stercorarius parasiticus*

英文名：Parasitic Jaeger

Cayenne Tern

STERNA CAYANA, Lath.

Male Adult, Spring plumage

Drawn from Nature by J. J. Audubon F.R.S. F.L.S.

Engraved, Printed & Coloured by R. Havell, London 1835

橙嘴凤头燕鸥（春季的成年雄性）

学　名：*Thalasseus maximus*

英文名：Royal Tern

Nº 55.

PLATE CCLXXIV.

Semipalmated Snipe, or Willet

Drawn from Nature by J. J. Audubon F.R.S. F.L.S.

TOTANUS SEMIPALMATUS, Temm. *Male Adult Spring plumage 1. Female Adult Winter plumage 2.*

Engraved, Printed & Coloured by R. Havell, London 1835

斑翅鹬

学　名：*Catoptrophorus semipalmatus*

英文名：Willet

1　春季的成年雄性　2　冬季的成年雌性

Noddy Tern

STERNA STOLIDA, L.

Male Adult

Drawn from Nature by J. J. Audubon F.R.S. F.L.S.

Engraved, Printed & Coloured by R. Havell, London 1835

白顶玄燕鸥（成年雄性）

学　名：*Anous stolidus*

英文名：Brown Noddy

No. 56.

PLATE CCLXXVI.

King Duck

FULIGULA SPECTABILIS, Lath.

Male 1. Female 2.

Drawn from Nature by J. J. Audubon F.R.S. F.L.S.

Engraved, Printed & Coloured by R. Havell, 1835

王绒鸭

学　名：*Somateria spectabilis*

英文名：King Eider

 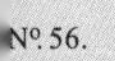

Hutchins's Barnacle Goose

Drawn from Nature by J. J. Audubon F.R.S. F.L.S. | ANSER HUTCHINSII, Richards. & Swain | Engraved, Printed & Coloured by R. Havell, 1835

加拿大黑雁

学　名：*Branta canadensis*

英文名：Canada Goose

No. 56.

PLATE CCLXXVIII.

Schinz's Sandpiper

TRINGA SCHINZII, Brehm

View on the East Coast of Florida

Drawn from Nature by J. J. Audubon F.R.S. F.L.S.

Engraved, Printed & Coloured by R. Havell, 1835

白腰滨鹬

学　名：*Calidris fuscicollis*

英文名：White-rumped Sandpiper

Sandwich Tern

STERNA BOYSSII, Lath.

Florida Cray Fish

Drawn from Nature by J. J. Audubon F.R.S. F.L.S.

Engraved, Printed & Coloured by R. Havell, 1835

白嘴端凤头燕鸥

学　名：*Thalasseus sandvicensis*

英文名：Sandwich Tern

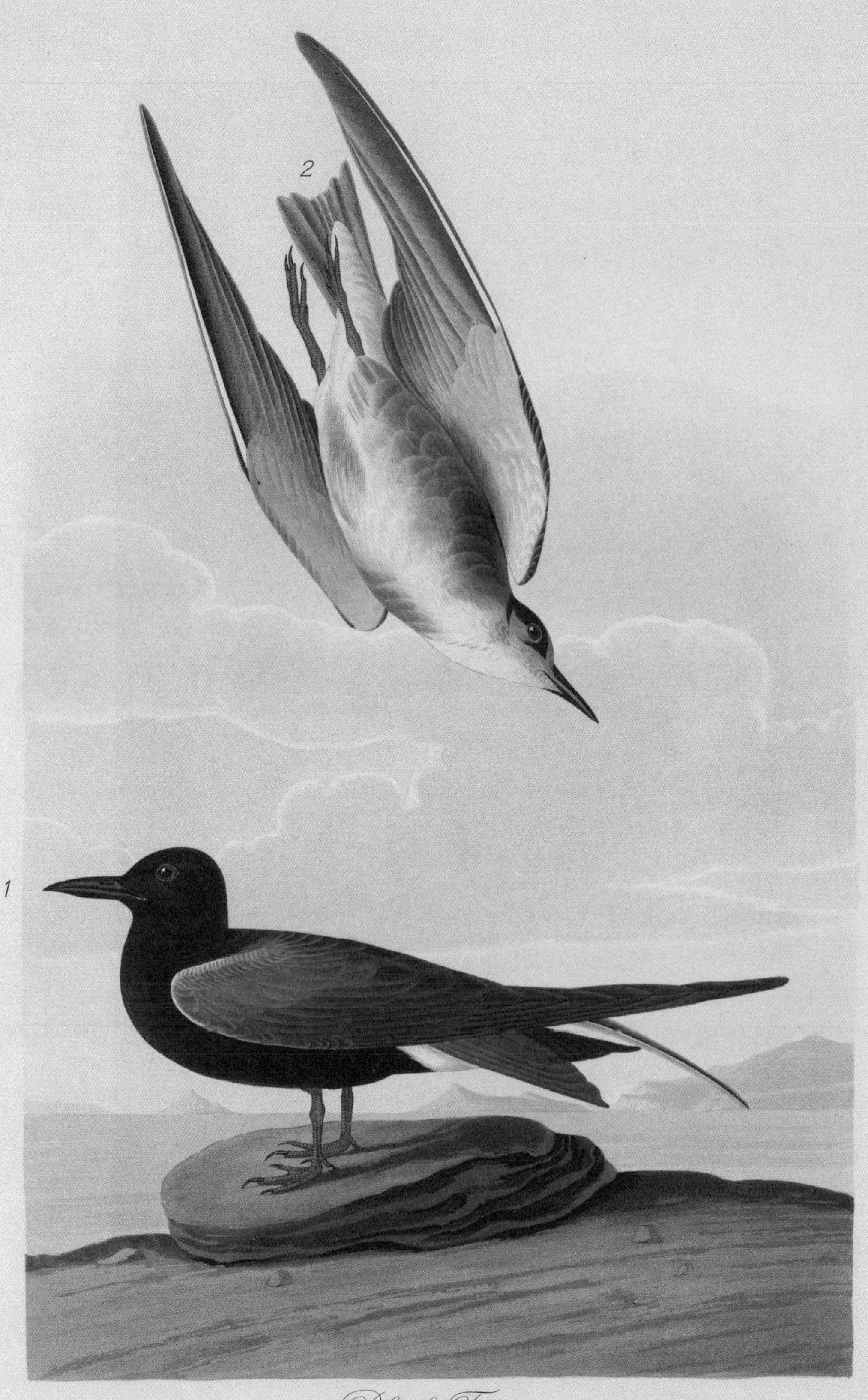

Black Tern

STERNA NIGRA, Linn.

Drawn from Nature by J. J. Audubon F.R.S. F.L.S.

Adult 1. Young in Autumn 2.

Engraved, Printed & Coloured by R. Havell, 1835

黑浮鸥

学　名：*Chlidonias niger*

英文名：Black Tern

1．成体　2．秋季的亚成体

Drawn from Nature by J. J. Audubon F.R.S. F.L.S.　*Great White Heron* ARDEA OCCIDENTALIS *Male adult spring plumage View Key-west*　Engraved, Printed & Coloured by R. Havell, 1835

大蓝鹭（白色型，春季的成年雄性）

学　名：*Ardea herodias*

英文名：Great Blue Heron

No. 57.

PLATE CCLXXXII.

Drawn from Nature by J. J. Audubon F.R.S. F.L.S. *White-winged silvery Gull* LARUS LEUCOPTERUS, Bonap *1. Male summer plumage 2. Young in Winter* Engraved, Printed & Coloured by R. Havell, 1835

冰岛鸥

学　名：*Larus glaucoides*

英文名：Iceland Gull

1．夏季的雄性　2．冬季的亚成体

Wandering Shearwater

Drawn from Nature by J. J. Audubon F.R.S. F.L.S.

PUFFINUS CINEREUS, Bonap *Male*

Engraved, Printed & Coloured by R. Havell, 1835

大　鹱（雄性）

学　名：*Ardenna gravis*

（原为 *Puffinus* 属）

英文名：Great Shearwater

Purple Sandpiper

TRINGA MARITIMA, Bonap

Drawn from Nature by J. J. Audubon F.R.S. F.L.S.

1. Male Summer plumage 2. Female Winter 3. Charadrius Wilsonius

Engraved, Printed & Coloured by R. Havell, 1835

紫滨鹬

学　名：*Calidris maritima*

英文名：Purple Sandpiper

厚嘴鸻（左 3）

学　名：*Charadrius wilsonia*

英文名：Wilson’s Plover

Fork-tailed Gull

LARUS SABINI, Swain & Richards

1. Male Summer plumage 2. Tringa arenaria Male Spring plumage

Drawn from Nature by J. J. Audubon F.R.S. F.L.S.

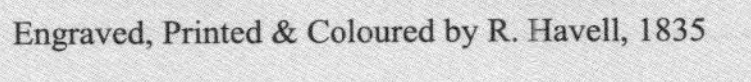
Engraved, Printed & Coloured by R. Havell, 1835

叉尾鸥

学　名：*Xema sabini*

英文名：Sabine’s Gull

1. 夏季的雄性　2. 三趾滨鹬，春季的雄性

White-fronted Goose Lath.

ANSER ALBIFRONS, Bechst

1. Male 2. Female

Drawn from Nature by J. J. Audubon F.R.S. F.L.S.

Engraved, Printed & Coloured by R. Havell, 1836

白额雁

学　名：*Anser albifrons*

英文名：Greater White-fronted Goose

1. 雄性　2. 雌性

Ivory Gull Lath.

Drawn from Nature by J. J. Audubon F.R.S. F.L.S.

LARUS EBURNEUS, Gm. *1. Adult Male 2. Young Second Autumn*

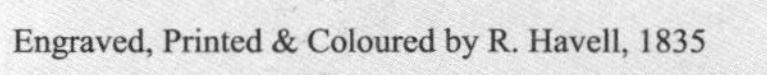
Engraved, Printed & Coloured by R. Havell, 1835

白 鸥

学　名：*Pagophila eburnea*

英文名：Ivory Gull

1. 成年雄性　2. 次年秋季的亚成体

Yellow Shank

TOTANUS FLAVIPES, Vieill

Male Summer plumage

View in South Carolina

小黄脚鹬（夏季的雄性）

学　名：*Tringa flavipes*

英文名：Lesser Yellowlegs

Drawn from Nature by J. J. Audubon F.R.S. F.L.S.

Engraved, Printed & Coloured by R. Havell, 1836.

Solitary Sandpiper

Drawn from Nature by J. J. Audubon F.R.S. F.L.S.

TOTANUS CHLOROPYGIUS, Vieill *1. Male 2. Female*

Engraved, Printed & Coloured by R. Havell, 1835

褐腰草鹬

学　名：*Tringa Solitaria*

英文名：Solitary Sandpiper

1．雄性　2．雌性

No. 58.

PLATE CCXC.

Red backed Sandpiper

Drawn from Nature by J. J. Audubon F.R.S. F.L.S.

TRINGA ALPINA, L. *1. Summer plumage 2. Winter plumage*

Engraved, Printed & Coloured by R. Havell, 1835

黑腹滨鹬

学　名：*Calidris alpina*

英文名：Dunlin

1　夏羽　2　冬羽

Herring Gull

LARUS ARGENTATUS, Brunn

1. Adult Male Spring plumage 2. Young in November

Raccoon Oysters & View of the entrance into St. Augustine

Drawn from Nature by J. J. Audubon F.R.S. F.L.S.　　Engraved, Printed & Coloured by R. Havell, 1836

银　鸥

学　名：*Larus argentatus*

英文名：European Herring Gull

1. 春季的成年雄性　2. 11 月的亚成体

No. 59.

PLATE CCXCII.

Crested Grebe

PODICEPS CRISTATUS, Lath.

Adult Male spring plumage 1. Young first Winter 2.

Drawn from Nature by J. J. Audubon F.R.S. F.L.S.

Engraved, Printed & Coloured by R. Havell, 1836

凤头䴙䴘

学　名：*Podiceps cristatus*

英文名：Great Crested Grebe

1. 春季的成年雄性　2. 第一年冬季的亚成体

1

2

Large billed Puffin

MORMON GLACIALIS, Leach

Drawn from Nature by J. J. Audubon F.R.S. F.L.S.

1. Male 2. Female

Engraved, Printed & Coloured by R. Havell, 1836

角海鹦

学　名：*Fratercula corniculata*

英文名：Horned Puffin

1．雄性　2．雌性

No. 59.

PLATE CCXCIV.

Pectoral Sandpiper

TRINGA PECTORALIS

1. Male 2. Female

Drawn from Nature by J. J. Audubon F.R.S. F.L.S.

Engraved, Printed & Coloured by R. Havell, 1836

斑胸滨鹬

学　名：*Calidris melanotos*

英文名：Pectoral Sandpiper

1. 雄性　2. 雌性

Drawn from Nature by J. J. Audubon F.R.S. F.L.S.

Manks Shearwater

Engraved, Printed & Coloured by R. Havell, 1836

PUFFINUS ANGLORUM, Ray

Male Bird

大西洋鸌（雄性）

学　名：*Puffinus puffinus*

英文名：Manx Shearwater

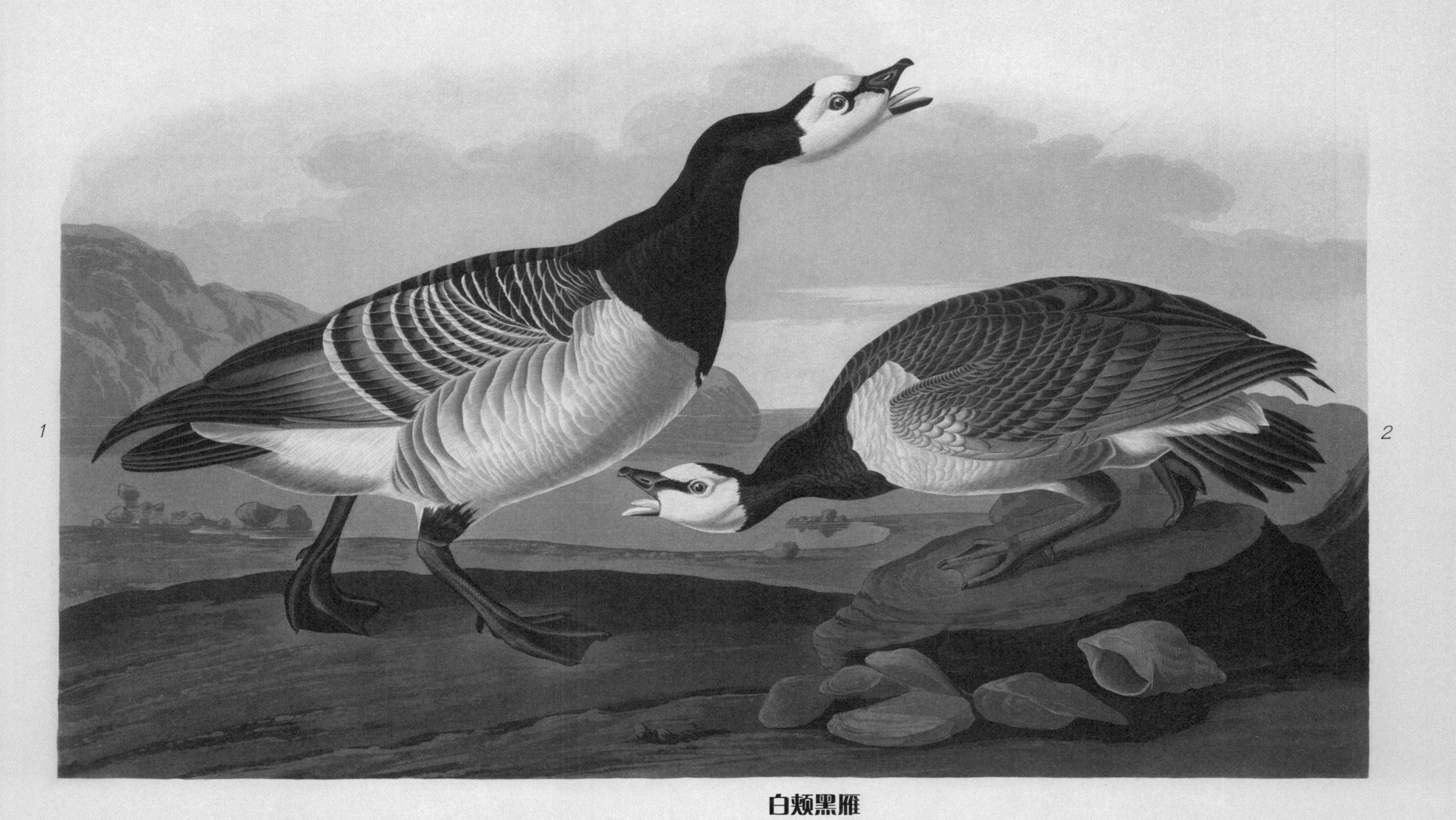

白颊黑雁

学　名：*Branta leucopsis*

英文名：Barnacle Goose

1. 雄性　2. 雌性

Harlequin Duck

FULIGULA HISTRIONICA, Bonap

1. Old Male 2. Female 3. Young Male third year

Drawn from Nature by J. J. Audubon F.R.S. F.L.S.

丑 鸭

学 名：*Histrionicus histrionicus*

英文名：Harlequin Duck

1. 老年雄性 2. 雌性 3. 3岁的雄性亚成体

Engraved, Printed & Coloured by R. Havell, 1836

Nº. 60.

PLATE CCXCVIII.

Red-necked Grebe

Drawn from Nature by J. J. Audubon F.R.S. F.L.S.

PODICEPS RUBRICOLLIS, Lath. *1. Adult Male spring plumage 2. Winter plumage*

Engraved, Printed & Coloured by R. Havell, 1836

赤颈䴙䴘

学　名：*Podiceps grisegena*

英文名：Red-necked Grebe

Dusky Petrel Lath.

PUFFINUS OBSCURUS, Cuvier

Male in Spring

Drawn from Nature by J. J. Audubon F.R.S. F.L.S.

Engraved, Printed & Coloured by R. Havell, 1836

奥氏鹱（春季的雄性）

学　名：*Puffinus lherminieri*

英文名：Audubon’s Shearwater

No. 60.

PLATE CCC.

2

3

1

Golden Plover

Drawn from Nature by J. J. Audubon F.R.S. F.L.S.

CHARADRIUS PLUVIALIS, L. *1. Summer plumage 2. Winter 3. Variety in March*

Engraved, Printed & Coloured by R. Havell, 1836

美洲金鸻

学　名：*Pluvialis dominica*

英文名：American Golden Plover

1

3

2

Canvas backed Duck

FULIGULA VALLISNERIA, Steph.

1, 2. Male 3. Female

View of Baltimore

Drawn from Nature by J. J. Audubon F.R.S. F.L.S.

Engraved, Printed & Coloured by R. Havell, 1836

帆背潜鸭

学　名：*Aythya valisineria*

英文名：Canvasback

1,2.　雄性　3.　雌性

Dusky Duck

Drawn from Nature by J. J. Audubon F.R.S. F.L.S. ANAS OBSCURA, Gm. 1 Male 2 Female Engraved, Printed & Coloured by R. Havell, 1836

北美黑鸭

学　名：*Anas rubripes*

英文名：American Black Duck

1　雄性　2　雌性

Bartram Sandpiper

TOTANUS BARTRAMIUS, Temm.

1. Male 2. Female

Drawn from Nature by J. J. Audubon F.R.S. F.L.S.

Engraved, Printed & Coloured by R. Havell, 1836

高原鹬

学　名：*Bartramia longicauda*

英文名：Upland Sandpiper

1. 雄性　2. 雌性

No. 61.

PLATE CCCIV.

Turn-stone

Drawn from Nature by J. J. Audubon F.R.S. F.L.S.

STREPSILAS INTERPRES, Ill *1. Summer plumage 2. Winter*

Engraved, Printed & Coloured by R. Havell, 1836

翻石鹬

学　名：*Arenaria interpres*

英文名：Ruddy Turnstone

1. 夏羽　2. 冬羽

Purple Gallinule

GALLINULA MARTINICA, Gmel.

Adult Male spring plumage

Drawn from Nature by J. J. Audubon F.R.S. F.L.S.

Engraved, Printed & Coloured by R. Havell, 1836

紫青水鸡（春季的成年雄性）

学　名：*Porphyrio martinicus*

英文名：Purple Gallinule

No. 62.

PLATE CCCVI.

Great Northern Diver or Loon

COLUMBUS GLACIALIS, L.

Adult 1. Young in Winter 2.

Drawn from Nature by J. J. Audubon F.R.S. F.L.S.

Engraved, Printed & Coloured by R. Havell, 1836

普通潜鸟

学　名：*Gavia immer*

英文名：Great Northern Loon

Blue Crane, or Heron

ARDEA CŒRULEA *1. Adult Male spring plumage 2. Young second year*

View near Charlestone S.C.

Drawn from Nature by J. J. Audubon F.R.S. F.L.S.

Engraved, Printed & Coloured by R. Havell, 1836

小蓝鹭

学　名：*Egretta caerulea*

英文名：Little Blue Heron

1. 春季的成年雄性　2. 次年的亚成体

No. 62.

PLATE CCCVIII.

Tell-tale Godwit or Snipe

Drawn from Nature by J. J. Audubon F.R.S. F.L.S.

TOTANUS MELANOLEUCUS, Vieill *Male 1. F. 2. View in East Florida*

Engraved, Printed & Coloured by R. Havell, 1836

大黄脚鹬

学　名：*Tringa melanoleuca*

英文名：Greater Yellowlegs

Great Tern

STERNA HIRUNDO, L.

Drawn from Nature by J. J. Audubon F.R.S. F.L.S.

Male Spring Plumage

Engraved, Printed & Coloured by R. Havell, 1836

普通燕鸥（春季的雄性）

学　名：*Sterna hirundo*

英文名：Common Tern

Nº 62.

PLATE CCCX.

Spotted Sandpiper

Drawn from Nature by J. J. Audubon F.R.S. F.L.S.

TOTANUS MACULARIUS *1. Adult Male 2. Female View on Bayou Sarah Louisiana*

Engraved, Printed & Coloured by R. Havell, 1836

斑腹矶鹬

学　名：*Actitis macularia*

英文名：Spotted Sandpiper

1　成年雄性　2．雌性

American White Pelican

PELICANUS AMERICANUS, Aud.

Drawn from Nature by J. J. Audubon F.R.S. F.L.S. *Male Adult* Engraved, Printed & Coloured by R. Havell, 1836

美洲鹈鹕（成年雄性）

学　名：*Pelecanus erythrorhynchos*

英文名：American White Pelican

No. 63.

PLATE CCCXII.

Long-tailed Duck

Drawn from Nature by J. J. Audubon F.R.S. F.L.S.

FULIGULA GLACIALIS *1. Male, summer plumage 2. Male winter 3. Female and young*

Engraved, Printed & Coloured by R. Havell, 1836

长尾鸭

学　名：*Clangula hyemalis*

英文名：Long-tailed Duck

1　夏季的雄性　2　冬季的雄性　3　雌性和雏鸟

Blue-Winged Teal

ANAS DISCORS, L.

Male 1. Female 2.

Drawn from Nature by J. J. Audubon F.R.S. F.L.S.

Engraved, Printed & Coloured by R. Havell, 1836

蓝翅鸭

学　名：*Anas discors*

英文名：Blue-winged Teal

1. 雄性　2. 雌性

Black-headed Gull

LARUS ATRICILLA, L.

Drawn from Nature by J. J. Audubon F.R.S. F.L.S.

Adult Male Spring Plumage 1. Young first Autumn 2.

Engraved, Printed & Coloured by R. Havell, 1836

笑 鸥

学　名：*Leucophaeus atricilla*

（原为 *Larus* 属）

英文名：Laughing Gull

Red-breasted Sandpiper

TRINGA ISLANDICA, L.

Drawn from Nature by J. J. Audubon F.R.S. F.L.S.

Engraved, Printed & Coloured by R. Havell, 1836

红腹滨鹬

学　名：*Calidris canutus*

英文名：Red Knot

No. 64. PLATE CCCXVI.

Black-bellied Darter

Drawn from Nature by J. J. Audubon F.R.S. F.L.S. PLOTUS ANHINGA, L. Engraved, Printed & Coloured by R. Havell, 1836

美洲蛇鹈

学　名：*Anhinga anhinga*

英文名：Anhinga

1

2

Black, or Surf Duck

Drawn from Nature by J. J. Audubon F.R.S. F.L.S.

FULIGULA PERSPICILLATA *Male Adult 1. Female 2.*

Engraved, Printed and Coloured by R. Havell, 1836

斑头海番鸭

学　名：*Melanitta perspicillata*

英文名：Surf Scoter

1. 雄性　2. 雌性

American Avocet

RECURVIROSTRA AMERICANA

Young in first Winter Plumage 1.

Adult 2.

褐胸反嘴鹬

学　名：*Recurvirostra americana*

英文名：American Avocet

1．第一年冬季的亚成体　2．成体

Drawn from Nature by J. J. Audubon F.R.S. F.L.S.

Engraved, Printed & Coloured by R. Havell. 1836

N°. 64.

PLATE CCCXIX.

Lesser Tern

STERNA MINUTA, L.

Drawn from Nature by J. J. Audubon F.R.S. F.L.S.

Adult Spring Plumage 1. Young in Sepr. 2.

Engraved, Printed & Coloured by R. Havell, 1836

小白额燕鸥

学　名：*Sternula antillarum*

英文名：Least Tern

1. 春季的成体　2. 亚成体

Little Sandpiper

Drawn from Nature by J. J. Audubon F.R.S. F.L.S.   TRINGA PUSILLA, Wils. *Male Adult Summer plumage 1. F. 2.*   Engraved, Printed & Coloured by R. Havell, 1836

美洲小滨鹬

学　名：*Calidris minutilla*

英文名：Least Sandpiper

Drawn from Nature by J. J. Audubon F.R.S. F.L.S.

Roseate Spoonbill

PLATALEA AJAJA, L.

Male Adult

粉红琵鹭（成年雄性）

学　名：*Platalea ajaja*

（原为 *Ajaja* 属）

英文名：Roseate spoonbill

Engraved, Printed and Coloured by R. Havell, 1836

No. 65.

PLATE CCCXXII.

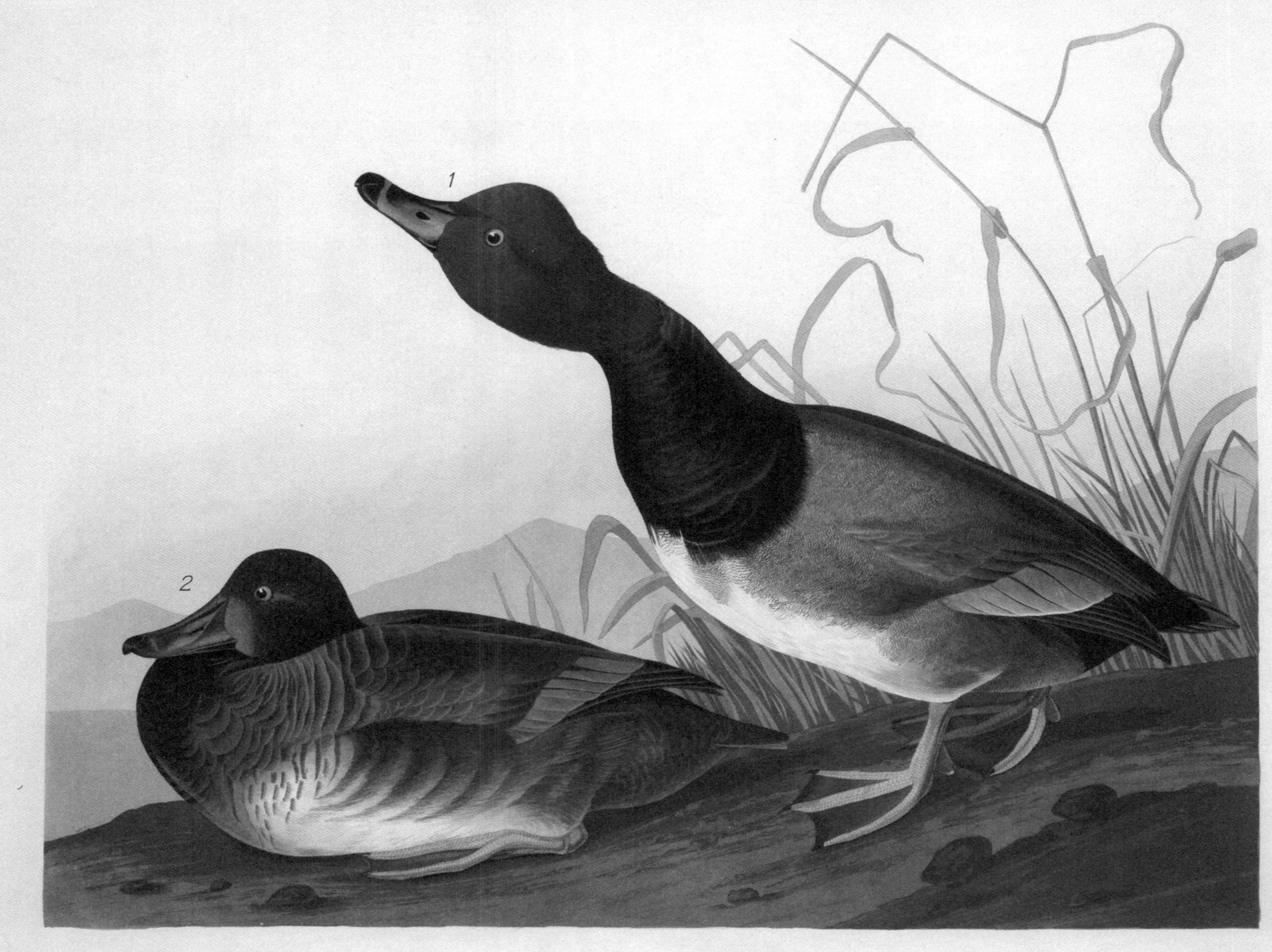

Red-headed Duck

FULIGULA FERINA, Steph.

Male 1. Female 2.

美洲潜鸭

学　名：*Aythya americana*

英文名：Redhead

1．雄性　2．雌性

Drawn from Nature by J. J. Audubon F.R.S. F.L.S.

Engraved, Printed & Coloured by R. Havell, 1836

Black Skimmer or Shearwater

RHINCOPS NIGRA, L.

rawn from Nature by J. J. Audubon F.R.S. F.L.S.　　*Male*　　Engraved, Printed & Coloured by R. Havell, 1836

黑剪嘴鸥（雄性）

学　名：*Rynchops niger*

英文名：Black Skimmer

Bonapartian Gull

LARUS BONAPARTII, Swain and Richards

Male Spring Plumage 1. Female 2.

Young first Autumn 3.

Drawn from Nature by J. J. Audubon F.R.S. F.L.S.

Engraved, Printed & Coloured by R. Havell, 183

博氏鸥

学　名：*Chroicocephalus philadelphia*

（原为 *Larus* 属）

英文名：Bonaparte's Gull

1. 春季的雄性　2. 雌性　3. 第一年秋季的亚成体

Buffel-headed Duck

FULIGULA ALBEOLA

Male 1. Female 2.

Drawn from Nature by J. J. Audubon F.R.S. F.L.S.

Engraved, Printed & Coloured by R. Havell, 1836

白枕鹊鸭

学　名：*Bucephala albeola*

英文名：Bufflehead

1. 雄性　2. 雌性

Nº 66.　　2　　1　　PLATE CCCXXVI.

Gannet

SULA BASSANA, Lacep.

Adult Male 1. Young first Winter 2.

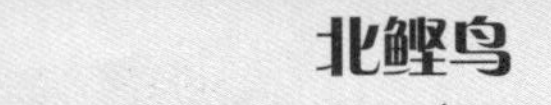

北鲣鸟

学　名：*Morus bassanus*

英文名：Northern Gannet

Shoveller Duck

ANAS CLYPEATA, L. *Male 1. Female 2.*

琵嘴鸭

学　名：*Anas clypeata*

英文名：Northern Shoveler

1. 雄性　2. 雌性

Drawn from Nature by J. J. Audubon F.R.S. F.L.S.

Engraved, Printed and Coloured by R. Havell, 1836

Long-legged Avocet

HIMANTOPUS NIGRICOLLIS, Vieill

Male

Drawn from Nature by J. J. Audubon F.R.S. F.L.S.

Engraved, Printed and Coloured by R. Havell, 1836

黑颈长脚鹬（雄性）

学　名：*Himantopus mexicanus*

英文名：Black-necked Stilt

Yellow-breasted Rail

RALLUS NOVEBORACENCIS, Bonap

Male Adult, Spring

Drawn from Nature by J. J. Audubon F.R.S. F.L.S.

Engraved, Printed and Coloured by R. Havell, 1836

北美花田鸡（春季的成年雄性）

学　名：*Coturnicops noveboracensis*

英文名：Yellow Rail

Ring Plover

Drawn from Nature by J. J. Audubon F.R.S. F.L.S.

Engraved, Printed and Coloured by R. Havell, 1836

CHARADRIUS SEMIPALMATUS

Adult Male 1. Young in August 2.

半蹼鸻

学　名：*Charadrius semipalmatus*

英文名：Semipalmated Plover

Drawn from Nature by J. J. Audubon F.R.S. F.L.S.

Goosander MERGUS MERGANSER, L. *Male 1. Female 2. Cohœs Falls, State of New York*

Engraved, Printed & Coloured by R. Havell, 1836

普通秋沙鸭

学　名：*Mergus merganser*

英文名：Common Merganser

1. 雄性　2. 雌性

N°.67.

PLATE CCCXXXII.

Pied Duck

Drawn from Nature by J. J. Audubon F.R.S. F.L.S.　　FULIGULA LABRADORA *Male Adult 1. Female 2.*　　Engraved, Printed and Coloured by R. Havell, 1836

拉布拉多鸭（已灭绝）

学　名：*Camptorhynchus labradorius*

英文名：Labrador Duck

1. 成年雄性　2. 雌性

Green Heron

Drawn from Nature by J. J. Audubon F.R.S. F.L.S.　　ARDEA VIRESCENS, L. *Adult Male 1. Young in September 2.*　　Engraved, Printed and Coloured by R. Havell, 1836

美洲绿鹭

学　名：*Butorides virescens*

英文名：Green Heron

1．成年雄性　2．9月的亚成体

No. 67.

PLATE CCCXXXIV.

Drawn from Nature by J. J. Audubon F.R.S. F.L.S.

Engraved, Printed and Coloured by R. Havell, 1836

Black-bellied Plover

CHARADRIUS HELVETICUS

Adult Male. Spring Plumage 1. Young in Autumn 2. Nestling 3.

灰　鸻

学　名：*Pluvialis squatarola*

英文名：Grey Plover

1. 春季的成年雄性　2. 秋季的亚成体　3. 雏鸟

Red-breasted Snipe

SCOLOPAX GRISEA, Gm.

Spring Plumage 1. Winter 2.

Drawn from Nature by J. J. Audubon F.R.S. F.L.S.

Engraved, Printed and Coloured by R. Havell, 1836

短嘴半蹼鹬

学　名：*Limnodromus griseus*

英文名：Short-billed Dowitcher

1. 春羽　2. 冬羽

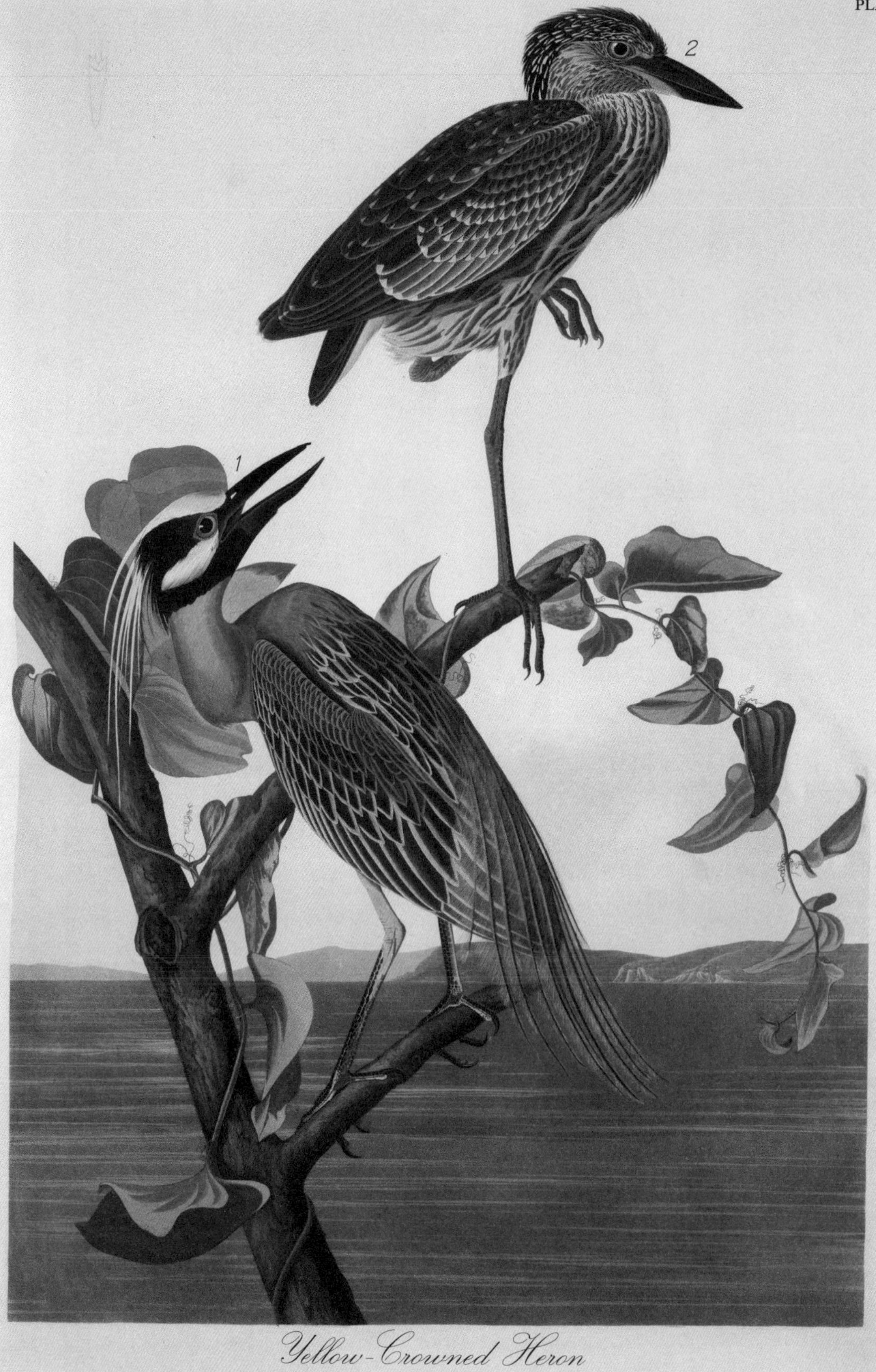

Yellow-Crowned Heron

ARDEA VIOLACEA, L.

Drawn from Nature by J. J. Audubon F.R.S. F.L.S.

Adult Male Spring Plumage 1. Young in October 2.

Engraved, Printed and Coloured by R. Havell, 1836

黄冠夜鹭

学　名：*Nyctanassa violacea*

英文名：Yellow-crowned Night Heron

1. 春季的成年雄性　2. 10月的亚成体

American Bittern

ARDEA MINOR *Male 1. Female 2.*

Drawn from Nature by J. J. Audubon F.R.S. F.L.S.

Engraved, Printed and Coloured by R. Havell, 1836

美洲麻鳽

学　名：*Botaurus lentiginosus*

英文名：American Bittern

1. 雄性　2. 雌性

No. 68. PLATE CCCXXXVIII.

Bemaculated Duck

Drawn from Nature by J. J. Audubon F.R.S. F.L.S.

ANAS GLOCITANS *Young Male in December*

Engraved, Printed and Coloured by R. Havell, 1836

布氏鸭（绿头鸭和赤膀鸭的杂交种，12月的雄性亚成体）

学　名：*Anas platyrhynchos × strepera*

英文名：Bemaculated Duck

Little Auk

Drawn from Nature by J. J. Audubon F.R.S. F.L.S.

URIA ALLE, Temm. *Male 1. Female 2.*

Engraved, Printed and Coloured by R. Havell, 1836

侏海雀

学　名：*Alle alle*

英文名：Little Auk

1. 雄性　2. 雌性

Nº. 68.

PLATE CCCXL.

Least Stormy-Petrel

THALASSIDROMA PELAGICA

Male 1. Female 2.

Drawn from Nature by J. J. Audubon F.R.S. F.L.S.

Engraved, Printed and Coloured by R. Havell, 1836

暴风海燕

学　名：*Hydrobates pelagicus*

英文名：European Storm Petrel

1　雄性　2　雌性

Great Auk

Drawn from Nature by J. J. Audubon F.R.S. F.L.S.

ALCA IMPENNIS, L.

Engraved, Printed and Coloured by R. Havell, 1836

大海雀（已灭绝）

学　名：*Pinguinus impennis*

英文名：Great Auk

1

2

Golden-Eye Duck

FULIGULA CLANGULA *Male 1. Female 2.*

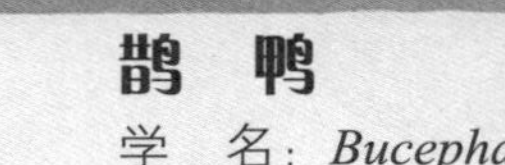

鹊　鸭

学　名：*Bucephala clangula*

英文名：Common Goldeneye

1. 雄性　2. 雌性

Drawn from Nature by J. J. Audubon F.R.S. F.L.S.

Engraved, Printed and Coloured by R. Havell, 1836

Ruddy Duck FULIGULA RUBIDA

Drawn from Nature by J. J. Audubon F.R.S. F.L.S.

Male adult 1. Female adult 2. Young Male second spring 3. Young 1st autumn 4.

Engraved, Printed & Coloured by R. Havell, 1836

棕硬尾鸭

学　名：*Oxyura jamaicensis*

英文名：Ruddy Duck

1. 成年雄性　2. 成年雌性　3. 第二年春季的雄性亚成体　4. 第一年秋季的亚成体

No. 69.

PLATE CCCXLIV.

Long-legged Sandpiper

Drawn from Nature by J. J. Audubon F.R.S. F.L.S.

TRINGA HIMANTOPUS

Engraved, Printed & Coloured by R. Havell, 1836

高跷鹬

学　名：*Calidris himantopus*

（原为 *Micropalama* 属）

英文名：Stilt Sandpiper

American Widgeon

Drawn from Nature by J. J. Audubon F.R.S. F.L.S.

ANAS AMERICANA, Gm. *Male 1. Female 2.*

Engraved, Printed & Coloured by R. Havell, 1836

绿眉鸭

学　名：*Anas Americana*

英文名：American Wigeon

1. 雄性　2. 雌性

No. 70. PLATE CCCXLVI.

Black-Throated Diver

Drawn from Nature by J. J. Audubon F.R.S. F.L.S.

COLYMBUS ARCTICUS, L. *Male 1. Female 2. Young in October 3.*

Engraved, Printed and Coloured by R. Havell, 1836

黑喉潜鸟

学　名：*Gavia arctica*

英文名：Black-throated Diver

1. 雄性　2. 雌性　3. 10 月的亚成体

Smew or White Nun

Drawn from Nature by J. J. Audubon F.R.S. F.L.S. MERGUS ALBELLUS, L. *Male 1. Female 2.* Engraved, Printed and Coloured by R. Havell, 1836

斑头秋沙鸭

学　名：*Mergellus albellus*

英文名：Smew

1. 雄性　2. 雌性

No. 70.

PLATE CCCXLVIII.

Gadwall Duck

ANAS STREPERA, L.

Male 1. Female 2.

Drawn from Nature by J. J. Audubon F.R.S. F.L.S.

Engraved, Printed and Coloured by R. Havell, 1836

赤膀鸭

学　名：*Anas strepera*

英文名：Gadwall

1．雄性　2．雌性

Least Water-hen, Edwards

RALLUS JAMAICENSIS, Gmel.

1. Male. 2. Young Adult

黑田鸡

学　名：*Laterallus jamaicensis*

英文名：Black Rail

1. 雄性　2. 亚成体

Drawn from Nature by J. J. Audubon F.R.S. F.L.S.

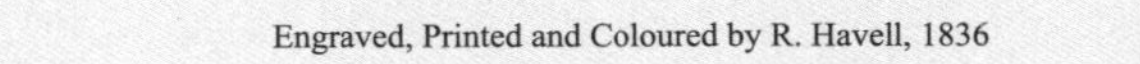

Engraved, Printed and Coloured by R. Havell, 1836

Nº 70. PLATE CCCL.

Rocky Mountain Plover

CHARADRIUS MONTANUS, Townsend

Adult Female

Drawn from Nature by J. J. Audubon F.R.S. F.L.S.

Engraved, Printed and Coloured by R. Havell, 1836

岩鸻（成年雌性）

学　名：*Charadrius montanus*

英文名：Mountain Plover

Great Cinereous Owl

STRIX CINEREA, —Gmelin

Drawn from Nature by J. J. Audubon F.R.S. F.L.S.

Female Adult

Engraved, Printed and Coloured by R. Havell, 1837

乌林鸮（成年雌性）

学　名：*Strix nebulosa*

英文名：Great Grey Owl

Black-Winged Hawk

FALCO DISPAR, Temm.

Drawn from Nature by J. J. Audubon F.R.S. F.L.S. *Male 1. Female 2.* Engraved, Printed and Coloured by R. Havell, 1837

黑翅鸢

学　名：*Elanus caeruleus*

英文名：Black-winged Kite

1. 雄性　2. 雌性

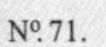

Chesnut-backed Titmouse
PARUS RUFESCENS, Townsend *1. Male 2. Female*

Black-capt Titmouse
PARUS ATRICAPILLUS, Wils. *3. Male 4. Female*
Willow Oak – Quercus Phellœs. L.

Chesnut-crowned Titmouse
PARUS MINIMUS, Townsend *5. Male 6. Female*
(and Nest)

Drawn from Nature by J. J. Audubon, F.R.S. F.L.S.

Engraved, Printed and Coloured by R. Havell 1837

1,2
栗背山雀
学　名：*Poecile rufescens*
（原为 *Parus* 属）
英文名：Chestnut-backed Chickadee
1. 雄性　2. 雌性

3,4
黑顶山雀
学　名：*Poecile atricapillus*
（原为 *Parus* 属）
英文名：Black-capped Chickadee
3. 雄性　4. 雌性

5,6
短嘴长尾山雀
学　名：*Psaltriparus minimus*
英文名：American Bushtit
5. 雄性　6. 雌性（和巢）

Louisiana Tanager

TANAGRA LUDOVICIANA, Wils.

1 & 2. Males Spring Plumage

Scarlet Tanager

TANAGRA RUBRA, L.

3. Old Male, Spring Plumage 4. Old Female. Do., Do.

Plant, Laurus Carolinensis

Drawn from Nature by J. J. Audubon F.R.S. F.L.S.

Engraved, Printed and Coloured by R. Havell, 1837

1,2

黄腹丽唐纳雀

学　名：*Piranga ludoviciana*

英文名：Western Tanager

1,2. 春季的雄性

3,4

猩红丽唐纳雀

学　名：*Piranga olivacea*

英文名：Scarlet Tanager

3. 春季的老年雄性　4. 老年雌性

PLATE CCCLV.

MacGillivray's Finch

FRINGILLA MACGILLIVRAII

Drawn from Nature by J. J. Audubon F.R.S. F.L.S.

Male 1. Female 2.

Engraved, Printed and Coloured by R. Havell, 1837

灰头地莺

学　名：*Oporornis tolmiei*

英文名：MacGillivray's Warbler

1. 雄性　2. 雌性

Marsh Hawk

FALCO CYANEUS

Drawn from Nature by J. J. Audubon F.R.S. F.L.S. *Male Adult 1. Female Adult 2. Young Male 3.* Engraved, Printed and Coloured by R. Havell, 1837

白尾鹞

学　名：*Circus cyaneus*

英文名：Hen Harrier

1．成年雄性　2．成年雌性　3．雄性，亚成体

American Magpie

CORVUS PICA

Drawn from Nature by J. J. Audubon F.R.S. F.L.S. *Male 1. Female 2.* Engraved, Printed and Coloured by R. Havell, 1837

黑嘴喜鹊

学　名：*Pica hudsonia*

英文名：Black-billed Magpie

1．雄性　2．雌性

Pine Grosbeak

PYRRHULA ENUCLEATOR

Drawn from Nature by J. J. Audubon F.R.S. F.L.S. *Male Adult, Spring Plumage 1. Female 2. Young first Winter 3.* Engraved, Printed and Coloured by R. Havell, 1837

松　雀

学　名：*Pinicola enucleator*

英文名：Pine Grosbeak

1. 春季的成年雄性　2. 雌性　3. 第一年冬季的亚成体

Arkansaw Flycatcher
MUSCICAPA VERTICALIS, Bonap
1. Male 2. Female

Swallow-Tailed Flycatcher
MUSCICAPA FORFICATA, Gme.
3. Male

Says Flycatcher
MUSCICAPA SAYA, Bonap
4. Male 5. Female

Drawn from Nature by J. J. Audubon F.R.S. F.L.S.

Engraved, Printed and Coloured by R. Havell, 1837

1,2

棕腹长尾霸鹟

学　名：*Sayornis saya*

英文名：Say’s Phoebe

1. 雄性　2. 雌性

3

剪尾王霸鹟

学　名：*Tyrannus forficatus*

英文名：Scissor-tailed Flycatcher

3. 雄性

4,5

西王霸鹟

学　名：*Tyrannus verticalis*

英文名：Western Kingbird

4. 雄性　5. 雌性

No. 72.
PLATE CCCLX.
3
4
1
2
Winter Wren
SYLVIA TROGLODYTES
Male 1. Female 2. Young in Autumn 3.
Rock Wren
TROGLODYTES OBSELATA, Say.
Female 4.
Drawn from Nature by J. J. Audubon F.R.S. F.L.S.
Engraved, Printed and Coloured by R. Havell, 1837

1,2,3

鹪 鹩

学 名：*Winter wren*

英文名：Troglodytes troglodytes

1. 雄性 2. 雌性 3. 秋季的亚成体

4

岩异鹩（雌性）

学 名：*Salpinctes obsoletus*

英文名：Rock Wren

Long-tailed or Dusky Grous

TETRAO OBSCURUS

Male 1. Female 2.

Drawn from Nature by J. J. Audubon F.R.S. F.L.S.

Engraved, Printed and Coloured by R. Havell, 1837

蓝镰翅鸡

学　名：*Dendragapus obscurus*

英文名：Dusky Grouse

1. 雄性　2. 雌性

No. 73. PLATE CCCLXII.

1 *Stellers Jay* — CORVUS STELLERII

2 *Yellow billed Magpie* — CORVUS NUTALLII, Aud.

3 *Ultramarine Jay* — CORVUS ULTRAMARINUS

4.5 *Clark's Crow* — CORVUS COLUMBIANUS, Wils.

Plant Platanus racemosus Nuttall
Acorn of Quercus macrocarpa. Mich

Drawn from Nature by J. J. Audubon F.R.S. F.L.S.

Engraved, Printed and Coloured by R. Havell, 1837

1

暗冠蓝鸦

学　名：*Cyanocitta stelleri*

英文名：Steller's Jay

2

黄嘴喜鹊

学　名：*Pica nuttalli*

英文名：Yellow-billed Magpie

3

西丛鸦

学　名：*Aphelocoma Calijornica*

英文名：Western Scrub-Jay

4, 5

北美星鸦

学　名：*Nucifraga columbiana*

英文名：Clark's Nutcracker

Bohemian Chatterer

BOMBYCILLA GARRULA *Male 1. Female 2.*

Drawn from Nature by J. J. Audubon F.R.S. F.L.S.

Pyrus Americanus Canadian Service Tree

Engraved, Printed & Coloured by R. Havell, 1837

太平鸟

学　名：*Bombycilla garrulus*

英文名：Bohemian Waxwing

1. 雄性　2. 雌性

White-winged Crossbill

LOXIA LEUCOPTERA, Gm.

Male adult 1, 2. Female adult 3. Young F. 4.

New Foundland alder

Drawn from Nature by J. J. Audubon F.R.S. F.L.S.

Engraved, Printed & Coloured by R. Havell, 1837

白翅交嘴雀

学　名：*Loxia leucoptera*

英文名：Two-barred Crossbill

1,2. 成年雄性　3. 成年雌性　4. 雌性，亚成体

Lapland Long-spur

FRINGILLA LAPONICA

Male Spring plumage 1. Male in Winter 2. Female 3.

Drawn from Nature by J. J. Audubon F.R.S. F.L.S.　　Engraved, Printed & Coloured by R. Havell, 1837

铁爪鹀

学　名：*Calcarius lapponicus*

英文名：Lapland Longspur

1. 春季的雄性　2. 冬季的雄性　3. 雌性

Iceland or Jer Falcon

FALCO ISLANDICUS, Lath.

Drawn from Nature by J. J. Audubon F.R.S. F.L.S.　　*Female Birds*　　Engraved, Printed & Coloured by R. Havell, 1837

矛　隼（雌性）

学　名：*Falco rusticolus*

英文名：Gyrfalcon

Band-tailed Pigeon

COLUMBA FASCIATA, Say

1. Male 2. Female

Plant Nuttall Cornel

Cornus Nuttalli. Aud.

Drawn from Nature by J. J. Audubon F.R.S. F.L.S.

Engraved, Printed and Coloured by R. Havell, 1837

斑尾鸽

学　名：*Patagioenas fasciata*

（原为 *Columba* 属）

英文名：Band-tailed Pigeon

1. 雄性　2. 雌性

No. 74. PLATE CCCLXVIII.

Rock Grous

TETRAO RUPESTRIS, Leach

Male in Winter 1. Female Summer Plummage 2. Young in August 3.

岩雷鸟

学　名：*Lagopus muta*

英文名：Rock Ptarmigan

1. 冬季的雄性　2. 夏季的雌性　3. 8月的亚成体

Nº. 74.	PLATE CCCLXIX.

1. *Mountain Mocking bird* Male
ORPHEUS MONTANUS, Townsend

Plant Mistletoe
Viscum Verticillatum

2.3. *Varied Thrush* Male & Female
TURDUS NÆVIUS, Gm.

Drawn from Nature by J. J. Audubon F.R.S. F.L.S.

Engraved, Printed & Coloured by R. Havell, 1837

1

高山弯嘴嘲鸫（雄性）

学　名：*Oreoscoptes montanus*

英文名：Sage Thrasher

2,3

杂色鸫（雄性和雌性）

学　名：*Ixoreus naevius*

英文名：Sage Thrasher

American Water Ouzel

Drawn from Nature by J. J. Audubon F.R.S. F.L.S.

CINCLUS AMERICANUS *Male 1. Female 2.*

Engraved, Printed & Coloured by R. Havell, 1837

美洲河乌

学　名：*Cinclus mexicanus*

英文名：American Dipper

1. 雄性　2. 雌性

Cock of the Plains

TETRAO UROPHASIANUS

Male 1. Female 2.

Drawn from Nature by J. J. Audubon F.R.S. F.L.S.

艾草松鸡

学　名：*Centrocercus urophasianus*

英文名：Sage Grouse

1. 雄性　2. 雌性

Engraved, Printed & Coloured by R. Havell, 1837

Common Buzzard

BUTEO VULGARIS

Female

Drawn from Nature by J. J. Audubon F.R.S. F.L.S.

Marsh Hare. Female, Lepus Palustris, Bachman

Engraved, Printed & Coloured by R. Havell, 1837

斯氏鵟（雌性）

学　名：*Buteo swainsoni*

英文名：Swainson’s Hawk

No. 75. PLATE CCCLXXIII.

Evening Grosbeak
FRINGILLA VESPERTINA, Cooper
Old Male 1.

Spotted Grosbeak
FRINGILLA MACULATA
Male 2, 3. Female 4.

Drawn from Nature by J. J. Audubon F.R.S. F.L.S.

Engraved, Printed & Coloured by R. Havell, 1837

1

黄昏锡嘴雀

学　名：*Hesperiphona vespertinus*

英文名：Evening Grosbeak

1. 老年雄性

2,3,4

黑头白斑翅雀

学　名：*Pheucticus melanocephalus*

英文名：Black-headed Grosbeak

2,3. 雄性　4. 雌性

Sharp-shinned Hawk

Drawn from Nature by J. J. Audubon F.R.S. F.L.S. FALCO VELOX, Wilson *Male 1. Female 2.* Engraved, Printed & Coloured by R. Havell, 1837

纹腹鹰

学　名：*Accipiter striatus*

英文名：Sharp-shinned Hawk

1. 雄性　2. 雌性

Lesser Red-Poll

FRINGILLA LINARIA, L.

Male 1. Female 2.

Plant Snow Berry

Symphoricarpes Racemosus

Drawn from Nature by J. J. Audubon F.R.S. F.L.S.

Engraved, Printed & Coloured by R. Havell, 1837

白腰朱顶雀

学　名：*Acanthis flammea*

英文名：Common Redpoll

1. 雄性　2. 雌性

No. 76.

PLATE CCCLXXVI.

Trumpeter Swan

CYGNUS BUCCINATOR *Young*

黑嘴天鹅（亚成体）

学　名：*Cygnus buccinator*

英文名：Trumpeter Swan

Drawn from Nature by J. J. Audubon F.R.S. F.L.S.

Engraved, Printed & Coloured by R. Havell, 1837

Scolopaceus Courlan

Drawn from Nature by J. J. Audubon F.R.S. F.L.S.

ARAMUS SCOLOPACEUS, Vieill

Engraved, Printed and Coloured by R. Havell, 1837

秧　鹤

学　名：*Aramus guarauna*

英文名：Limpkin

N°.76. PLATE CCCLXXVIII.

Hawk Owl

STRIX FUNEREA

Drawn from Nature by J. J. Audubon F.R.S. F.L.S. *Male 1. Female 2.* Engraved, Printed and Coloured by R. Havell, 1837

猛 鸮

学　名：*Surnia ulula*

英文名：Northern Hawk-owl

1. 雄性　2. 雌性

Ruff-necked Humming-bird

TROCHILUS RUFUS, Latham

1, 2. Males 3. Female & Nest

Drawn from Nature by J. J. Audubon F.R.S. F.L.S.

Plant Cleome heptaphylla

Engraved, Printed & Coloured by R. Havell, 1837

棕煌蜂鸟

学　名：*Selasphorus rufus*

英文名：Rufous Hummingbird

1,2. 雄性　3. 雌性和巢

Tengmalm's Owl

STRIX TENGMALMI

Drawn from Nature by J. J. Audubon F.R.S. F.L.S.

Male 1. Female 2.

Engraved, Printed and Coloured by R. Havell, 1837

鬼　鸮

学　名：*Aegolius funereus*

英文名：Boreal Owl

1. 雄性　2. 雌性

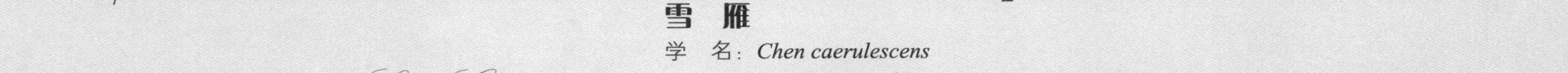
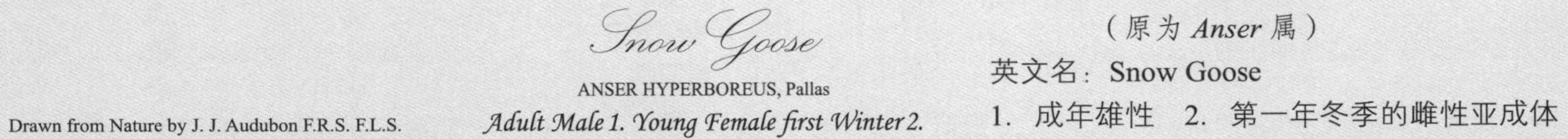
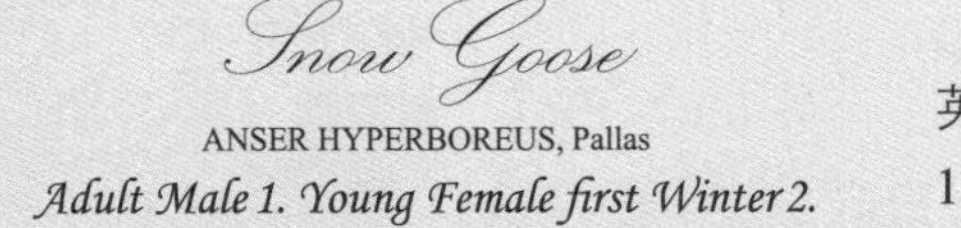

Snow Goose

ANSER HYPERBOREUS, Pallas

Adult Male 1. Young Female first Winter 2.

Drawn from Nature by J. J. Audubon F.R.S. F.L.S.

Engraved, Printed and Coloured by R. Havell, 1837.

雪　雁

学　名：*Chen caerulescens*

（原为 *Anser* 属）

英文名：Snow Goose

1. 成年雄性　2. 第一年冬季的雌性亚成体

2 1

Sharp-tailed Grous

Drawn from Nature by J. J. Audubon F.R.S. F.L.S.

TETRAO PHASIANELLUS *Male 1. Female 2.*

Engraved, Printed and Coloured by R. Havell, 1837

尖尾松鸡

学　名：*Tympanuchus phasianellus*

英文名：Sharp-tailed Grouse

1. 雄性　2. 雌性

Long-eared Owl

STRIX OTUS

Male

Drawn from Nature by J. J. Audubon F.R.S. F.L.S.

Engraved, Printed & Coloured by R. Havell, 1837

长耳鸮（雄性）

学　名：*Asio otus*

英文名：Long-eared Owl

Black-throated Bunting

FRINGILLA AMERICANA

Drawn from Nature by J. J. Audubon F.R.S. F.L.S. *Male 1. Female 2.* Engraved, Printed & Coloured by R. Havell, 1837

美洲雀

学　名：*Spiza americana*

英文名：Dickcissel

1. 雄性　2. 雌性

Bank Swallow

HIRUNDO RIPARIA

Male 1. Female 2. Young 3.

Violet-Green Swallow

HIRUNDO THALASSINUS, Swain

Male 4. Female 5.

Drawn from Nature by J. J. Audubon F.R.S. F.L.S.

Engraved, Printed & Coloured by R. Havell, 1837

1,2,3

崖沙燕

学　名：*Riparia riparia*

英文名：Sand Martin

1. 雄性　2. 雌性　3. 亚成体

4,5

紫绿树燕

学　名：*Tachycineta thalassina*

英文名：Violet-green Swallow

4. 雄性　5. 雌性

Nº 78.

PLATE CCCLXXXVI.

大白鹭（春季的雄性）

学　名：*Ardea alba*

（原为 *Egretta* 属）

英文名：Great Egret

White Heron

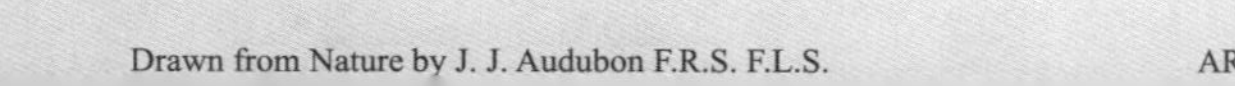

Drawn from Nature by J. J. Audubon F.R.S. F.L.S.

ARDEA ALBA, Linn. *Male, Spring Plumage*

Engraved, Printed & Coloured by R. Havell, 1837

Glossy Ibis

IBIS FALCINELLUS *Male Adult*

彩　鹮（成年雄性）

学　名：*Plegadis falcinellus*

英文名：Glossy Ibis

Drawn from Nature by J. J. Audubon F.R.S. F.L.S.

Engraved, Printed and Coloured by R. Havell, 1837

Nuttall's Starling

ICTERUS TRICOLOR, Aud.

1. Adult Male

Yellow-headed Troopial

ICTERUS XANTHOCEPHALUS, Bonap

2. Adult Male 3. Do. Female 4. head of Young Male

Bullock's Oriole

ICTERUS BULLOCKII

5. Adult Male

Drawn from Nature by J. J. Audubon F.R.S. F.L.S.

Engraved, Printed & Coloured by R. Havell, 1837

1

三色黑鹂（成年雄性）

学　名：*Agelaius tricolor*

英文名：Tricolored Blackbird

2, 3, 4

黄头黑鹂

学　名：*Xanthocephalus xanthocephalus*

英文名：Yellow-headed Blackbird

2. 成年雄性　3. 雌性　4. 雄性，亚成体，头部

5

布氏拟鹂（成年雄性）

学　名：*Icterus bullockii*

英文名：Bullock's Oriole

Red-Cockaded Woodpecker

PICUS QUERULUS, Wils.

Drawn from Nature by J. J. Audubon F.R.S. F.L.S.

Males 1. Female 2.

Engraved, Printed & Coloured by R. Havell, 1837

红顶啄木鸟

学　名：*Picoides borealis*

英文名：Red-cockaded Woodpecker

1. 雄性　2. 雌性

Lark Finch
FRINGILLA GRAMMACA, Say
1. Male

Prairie Finch
FRINGILLA BICOLOR, Townsend
2. Male 3. Female

Brown Song Sparrow
FRINGILLA CINEREA, Gmel.
4. Male

Drawn from Nature by J. J. Audubon F.R.S. F.L.S.

Engraved, Printed & Coloured by R. Havell, 1837

1

鹨雀鹀（雄性）

学　名：*Chondestes grammacus*

英文名：Lark Sparrow

2, 3

白斑黑鹀

学　名：*Calamospiza melanocorys*

英文名：Lark Bunting

2. 雄性　3. 雌性

4

歌带鹀（雄性）

学　名：*Melospiza melodia*

英文名：Song Sparrow

Brant Goose

ANSER BERNICLA

1. Male 2. Female

Drawn from Nature by J. J. Audubon F.R.S. F.L.S.

Engraved, Printed and Coloured by R. Havell, 1837

黑　雁

学　名：*Branta bernicla*

英文名：Brent Goose

1. 雄性　2. 雌性

Nº 79.

PLATE CCCXCII.

Louisiana Hawk

BUTEO HARRISI, Aud.

Adult Female

Drawn from Nature by J. J. Audubon F.R.S. F.L.S.

Engraved, Printed and Coloured by R. Havell, 1837

栗翅鹰（成年雌性）

学　名：*Parabuteo unicinctus*

英文名：Harris's Hawk

Drawn from Nature by J. J. Audubon F.R.S. F.L.S.

Engraved, Printed & Coloured by R. Havell, 1837

Townsend's Warbler

SYLVIA TOWNSENDI, Nuttall

1. Male

Arctic Blue-bird

SIALIA ARCTICA, Swain

2. Male 3. Female

Western Blue-bird

SIALIA OCCIDENTALIS, Townsend

4. Male 5. Female

Plant { *Carolina Allspice* / CALYCANTHUS FLORIDUS }

1

黄眉林莺（雄性）

学　名：*Setophaga townsendi*

（原为 *Dendroica* 属）

英文名：Townsend's Warbler

2,3

山蓝鸲

学　名：*Sialia currucoides*

英文名：Mountain Bluebird

2. 雄性　3. 雌性

4,5

西蓝鸲

学　名：*Sialia mexicana*

英文名：Western Bluebird

4. 雄性　5. 雌性

Drawn from Nature by J. J. Audubon F.R.S. F.L.S.

Engraved, Printed & Coloured by R. Havell, 1837

Chestnut-coloured Finch

PLECTROPHANES ORNATA, Towns

1. Male Spring

Black-headed Siskin

FRINGILLA MAGELLANICA, Vieill

2. Old Male

Black crown Bunting, Lath.

EMBERIZA ATRICAPILLA, Gmel.

3. Adult Male

1

栗领铁爪鹀（春季的雄性）

学　名：*Calcarius ornatus*

英文名：Chestnut-collared Longspur

2

黑头白斑翅雀（老年雄性）

学　名：*Pheucticus melanocephalus*

英文名：Black-headed grosbeak

3

金冠带鹀（成年雄性）

学　名：*Zonotrichia atricapilla*

英文名：Golden-crowned Sparrow

Arctic Ground Finch

PIPILO ARCTICA, Swain

4. Male 5. Female

4,5

棕胁唧鹀

学　名：*Spizella pallida*

英文名：Clay-colored Sparrow

4. 雄性　5. 雌性

No. 79. PLATE CCCXCV.

Drawn from Nature by J. J. Audubon F.R.S. F.L.S.

Engraved, Printed & Coloured by R. Havell, 1837

Audubon's Warbler

SYLVIA AUDUBONI, Townsend

1. Male 2. Female

Hermit Warbler

SYLVIA OCCIDENTALIS, Townsend

3. Male 4. Female

Plant. Strawberry Tree

EUYONUMUS AMERICANA

Black-throated gray Warbler

SYLVIA NIGRESCENS, Townsend

5 and 6. Males

1, 2

黄腰林莺

学　名：*Steophaga coronata*

英文名：Yellow-rumped Warbler

1. 雄性　2. 雌性

3, 4

黄脸林莺

学　名：*Dendroica occidentalis*

英文名：Hermit Warbler

3. 雄性　4. 雌性

5, 6

黑喉灰林莺

学　名：*Dendroica nigrescens*

英文名：Black-throated Gray warbler

5,6. 雄性

No. 80. PLATE CCCXCVI.

Burgomaster Gull

Drawn from Nature by J. J. Audubon F.R.S. F.L.S. LARUS GLAUCUS, Brunnich *1. Adult Male 2. Young, first Autumn* Engraved, Printed and Coloured by R. Havell, 1837

北极鸥

学　名：*Larus hyperboreus*

英文名：Glaucous Gull

1. 成年雄性　2. 第一年秋季的亚成体

1　　Scarlet Ibis　　2

IBIS RUBRA, Vieill

1. Adult Male 2. Young Second Autumn

Drawn from Nature by J. J. Audubon F.R.S. F.L.S.

Engraved, Printed and Coloured by R. Havell 1837

美洲红鹮

学　名：*Eudocimus ruber*

英文名：Scarlet Ibis

1. 成年雄性　2. 第二年秋季的亚成体

Drawn from Nature by J. J. Audubon F.R.S. F.L.S.

Engraved, Printed & Coloured by R. Havell, 1837

Lazuli Finch

FRINGILLA AMŒNA

1. Male Spring Plumamge

Clay-Coloured Finch

FRINGILLA PALLIDA, Swains

2. Male

Plant Liberty Bush

AZALEA NUDIFLORA

Oregon Snow Finch

FRINGILLA OREGONA, Towns

3. Male 4. Female

1

白腹蓝彩鹀（春季的雄性）

学　名：*Passerina amoena*

英文名：Lazuli Bunting

2

褐雀鹀（雄性）

学　名：*Spizella pallida*

英文名：Clay-colored Sparrow

3,4

暗眼灯草鹀

学　名：*Junco hyemalis*

英文名：Dark-eyed Junco

3. 雄性　4. 雌性

Drawn from Nature by J. J. Audubon F.R.S. F.L.S.

Engraved, Printed & Coloured by R. Havell, 1837

Black-throated green Warbler

SYLVIA VIRENS

1. Male 2. Female

Blackburnian w.

SYLVIA BLACKBURNIÆ

3. Female

Mourning Warbler

SYLVIA PHILADELPHIA

4. Male 5. Female

1, 2

黑喉绿林莺

学　名：*Setophaga virens*

（原为 *Dendroica* 属）

英文名：Black-throated Green Warbler

1. 雄性　2. 雌性

3

橙胸林莺（雌性）

学　名：*Setophaga fusca*

（原为 *Dendroica* 属）

英文名：Blackburnian Warbler

4, 5

灰头地莺

学　名：*Geothlypis tolmiei*

（原为 *Oporornis* 属）

英文名：MacGillivray's Warbler

4. 雄性　5. 雌性

Drawn from Nature by J. J. Audubon F.R.S. F.L.S.　Engraved, Printed & Coloured by R. Havell, 1837

Arkansaw Siskin

FRINGILLA SPALTRIA

1. Male

1

暗背金翅雀（雄性）

学　名：*Carduelis psaltria*

英文名：Dark-backed Goldfinch

Mealy Red-poll

LINOTA BOREALIS

2. Male

2

白腰朱顶雀（雄性）

学　名：*Acanthis flammea*

英文名：Common Redpoll

Louisiana Tanager

TANAGRA LUDOVICIANA

3. Female

3

黄腹丽唐纳雀（雌性）

学　名：*Piranga ludoviciana*

英文名：Western Tanager

Townsend's Finch

EMBERIZA TOWNSENDI

4. Male

4

唐氏雀（雄性）

学　名："*Spiza townsendi*"

英文名：Townsend's Finch

Buff-breasted Finch

EMBERIZA PICTA

5. Male

5

黄腹铁爪鹀（雄性）

学　名：*Calcarius pictus*

英文名：Smith's Longspur

2

1

Red-breasted Merganser

MERGUS SERRATOR, L.

Male 1. Female 2.

Plant Sarracenia flava

Drawn from Nature by J. J. Audubon F.R.S. F.L.S.

Engraved, Printed and Coloured by Rob. Havell, 1838

红胸秋沙鸭

学　名：*Mergus serrator*

英文名：Red-breasted Merganser

1. 雄性　2. 雌性

No. 81.

PLATE CCCCII.

Drawn from Nature by J. J. Audubon F.R.S. F.L.S.

Engraved, Printed and Coloured by Robt. Havell, 1838

Black-throated Guillemot

MERGULUS ANTIQUUS, Bonap

1, 2 *1. Adult 2. Young*

扁嘴海雀

学　名：*Synthliboramphus antiquus*

英文名：Ancient Murrelet

1. 成体　2. 亚成体

Nobbed-billed Auk

PHALERIS NODIROSTRIS, Bonap

3

小海雀

学　名：*Aethia pusilla*

英文名：Least Auklet

Curled-Crested Auk

PHALERIS SUPERCILIATA, Bonap

4

凤头海雀

学　名：*Aethia cristatella*

英文名：Crested Auklet

Horned-billed Guillemot

CERATORRHINA OCCIDENTALIS, Bonap

5

角嘴海雀

学　名：*Cerorhinca monocerata*

英文名：Rhinoceros Auklet

Golden-eye Duck

CLANGULA VULGARIS

Male, Summer Plumage

Drawn from Nature by J. J. Audubon F.R.S. F.L.S.

Engraved, Printed & Coloured by Robt. Havell, 1838

巴氏鹊鸭（夏季的雄性）

学　名：*Bucephala islandica*

英文名：Barrow’s Goldeneye

1

2

Eared Grebe

PODICEPS AURITUS

1. Adult 2. Young first Winter

Drawn from Nature by J. J. Audubon F.R.S. F.L.S.

Engraved, Printed & Coloured by Robt. Havell, 1838

黑颈䴙䴘

学　名：*Podiceps nigricollis*

英文名：Black-necked Grebe

1. 成体，繁殖羽　2. 第一年冬季的亚成体，非繁殖羽

Semipalmated Sandpiper

Drawn from Nature by J. J. Audubon F.R.S. F.L.S.

TRINGA SEMIPALMATA, Wils.

Engraved, Printed & Coloured by Robt. Havell, 1838

半蹼滨鹬

学　名：*Calidris pusilla*

英文名：Semipalmated Sandpiper

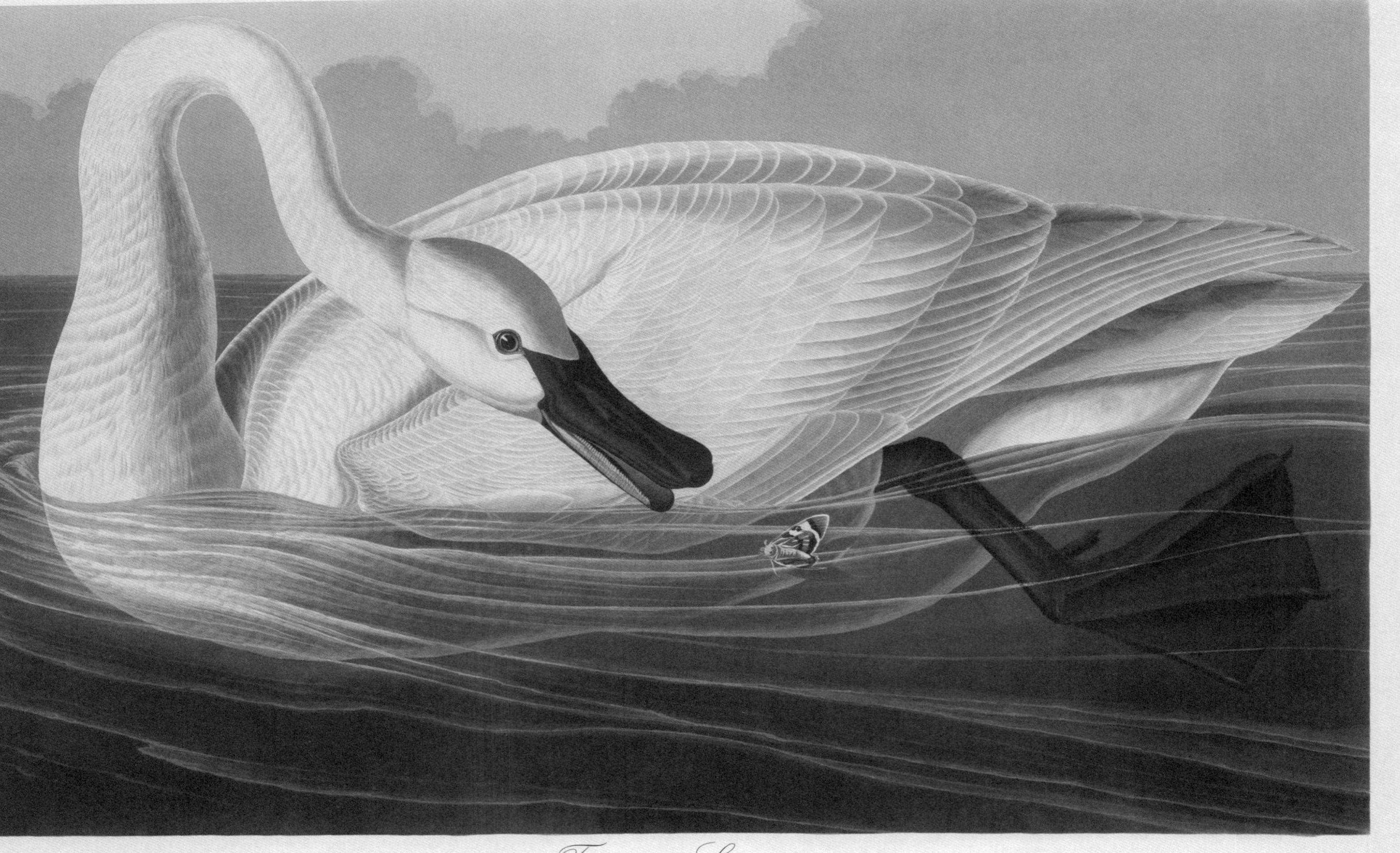

Trumpeter Swan

Drawn from Nature by J. J. Audubon F.R.S. F.L.S.

CYGNUS BUCCINATOR, Richardson

Adult

Engraved, Printed & Coloured by Robt. Havell, 1838

黑嘴天鹅（成体）

学　名：*Cygnus buccinator*

英文名：Trumpeter Swan

Dusky Albatros

DIOMEDEA FUSCA

Drawn from Nature by J. J. Audubon F.R.S. F.L.S.

Engraved, Printed and Coloured by Robt. Havell, 1838

乌信天翁

学　名：*Phoebetria fusca*

英文名：Sooty Albatross

Nº 82.

PLATE CCCCVIII.

American Scoter Duck

FULIGULA AMERICANA

Male 1. Female 2.

Drawn from Nature by J. J. Audubon F.R.S. F.L.S.

Engraved, Printed and Coloured by Robt. Havell, 1838

黑海番鸭

学　名：*Melanitta americana*

（原为 *nigra* 属）

英文名：Black Scoter

1. 雄性　2. 雌性

Havell's Tern, 1.

Trudeau's Tern, 2.

Drawn from Nature by J. J. Audubon F.R.S. F.L.S.

STERNA HAVELLI, Aud.

STERNA TRUDEAUI, Aud.

Engraved, Printed and Coloured by Robt. Havell, 1838

1

弗氏燕鸥

学　名：*Sterna forsteri*

英文名：Forster's Tern

2

白顶燕鸥

学　名：*Sterna trudeaui*

英文名：Snowy-crowned Tern

Marsh Tern

Drawn from Nature by J. J. Audubon F.R.S. F.L.S. STERNA ANGLICA, Montagu *Male, Summer Plumage* Engraved, Printed and Coloured by Robt. Havell, 1838

鸥嘴噪鸥（夏季的雄性）

学　名：*Gelochelidon nilotica*

英文名：Gull-billed Tern

Common American Swan

CYGNUS AMERICANUS, Sharplefs

Drawn from Nature by J. J. Audubon F.R.S. F.L.S.

Nymphea flava - Leitner

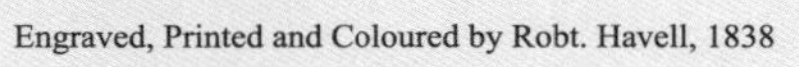

Engraved, Printed and Coloured by Robt. Havell, 1838

小天鹅（亚成体）

学　名：*Cygnus columbianus*

英文名：Tundra Swan

Violet-green Cormorant

PHALACROCORAX RESPLENDENS, Aud.

Female in Winter

Townsend's Cormorant

PHALACROCORAX TOWNSENDI, Aud.

Male

Drawn from Nature by J. J. Audubon F.R.S. F.L.S.

Engraved, Printed and Coloured by Robt. Havell, 1838

海鸬鹚（冬季的雌性）

学　名：*Phalacrocorax pelagicus*

英文名：Pelagic Cormorant

加州鸬鹚（雄性）

学　名：*Phalacrocorax penicillattus*

英文名：Brandt's Cormorant

Californian Partridge

PERDIX CALIFORNICA, Lath.

Male 1. Female 2.

Drawn from Nature by J. J. Audubon F.R.S. F.L.S.

Engraved, Printed & Coloured by Robt. Havell, 1838

珠颈斑鹑

学　名：*Callipepla californica*

英文名：California Quail

1. 雄性　2. 雌性

No. 83. PLATE CCCCXIV.

SYLVIA CHRYSOPTERA, Lath. SYLVIA MARITIMA, Wils.

Drawn from Nature by J. J. Audubon F.R.S. F.L.S. *Male 1. Female 2.* *Male 3. Female 4.* Engraved, Printed & Coloured by Robt. Havell, 1838

1, 2

金翅虫森莺

学　名：*Vermivora chrysoptera*

英文名：Golden-winged Warbler

1. 雄性　2. 雌性

3, 4

栗颊林莺

学　名：*Setophaga tigrina*

（原为 *Dendroica* 属）

英文名：Cape May Warbler

3. 雄性　4. 雌性

Brown Creeper *Californian Nuthatch*

CERTHIA FAMILIARIS, Linn. SITTA PYGMEA, Vig.

Drawn from Nature by J. J. Audubon F.R.S. F.L.S. *Male 1. Female 2.* *3, 4.* Engraved, Printed & Coloured by Robt. Havell, 1838

1, 2

美洲旋木雀

学　名：*Certhia americana*

英文名：Brown Creeper

1. 雄性　2. 雌性

3, 4

褐头䴓

学　名：*Sitta pusilla*

英文名：Brown-headed Nuthatch

Drawn from Nature by J. J. Audubon F.R.S. F.L.S.

Engraved, Printed & Coloured by Robt. Havell, 1838

Hairy Woodpecker

PICUS VILLOSUS, Linn.

1. Male 2. Female

Red-bellied Woodpecker

PICUS CAROLINUS, Linn.

3. Male 4. Female

Red-shafted Woodpecker

PICUS MEXICANUS, Aud.

5. Male 6. Female

1,2

长嘴啄木鸟

学　名：*Picoides villosus*

英文名：Hairy Woodpecker

1. 雄性　2. 雌性

3,4

红腹啄木鸟

学　名：*Melanerpes carolinus*

英文名：Red-bellied Woodpecker

3. 雄性　4. 雌性

5,6

北扑翅䴕

学　名：*Colaptes auratus*

英文名：Northern Flicker

5. 雄性　6. 雌性

Lewis' Woodpecker

PICUS TORQUATUS, Wils.

7. Male 8. Female

7,8

刘氏啄木鸟

学　名：*Melanerpes lewis*

英文名：Lewis's woodpecker

7. 雄性　8. 雌性

Red-breasted Woodpecker

PICUS RUBER, Lath.

9. Male 10. Female

9,10

红胸吸汁啄木鸟

学　名：*Sphyrapicns ruber*

英文名：Red-brested Sapsucker

9. 雄性　10. 雌性

Drawn from Nature by J. J. Audubon F.R.S. F.L.S.

Engraved, Printed and Coloured by Robt. Havell, 1838

Maria's Woodpecker
PICUS MARTINI, Aud.
1. Male 2. Female

Three-toed Woodpecker
PICUS HIRSITUS, Vieill
3. Male 4. Female

Phillips' Woodpecker
PICUS PHILLIPSI, Aud.
5 and 6. Males

Canadian Woodpecker
PICUS CANADENSIS, Buff.
7. Male

Harris's Woodpecker
PICUS HARRISI, Aud.
8. Male 9. Female

Audubon's Woodpecker
PICUS AUDUBONI, Trudeau
10. Male

1, 7, 8（雄性）
2, 9（雌性）

长嘴啄木鸟

学　名：*Picoides villosus*

英文名：Hairy woodpecker

3, 5, 6, 10（雄性）
4（雌性）

三趾啄木鸟（美洲）

学　名：*Picoides dorsalus*

英文名：American Three-toed Woodpecker

Nº. 84.

PLATE CCCCXVIII.

American Ptarmigan

TETRAO MUTUS, Leach

Male Spring Plumamge

White-tailed Grous

TETRAO LEUCURUS, Swains

Winter Plumage

Drawn from Nature by J. J. Audubon F.R.S. F.L.S.

Engraved, Printed & Coloured by Robt. Havell, 1838

柳雷鸟（左，春季的雄性）

学　名：*Lagopus lagopus*

英文名：Willow Grouse

白尾雷鸟（右，冬季羽征）

学　名：*Lagopus Leucura*

英文名：White-tailed Ptarmigan

Drawn from Nature by J. J. Audubon F.R.S. F.L.S.

Engraved, Printed & Coloured by Robt. Havell, 1838

Little Tawny Thrush
TURDUS MINOR, Gm.
1. Male

Ptiliogony's Townsendi Aud
2. Female

Canada Jay
CORVUS CANADENSIS, Linn.
3. Young, Male

1
隐夜鸫（雄性）
学　名：*Catharus guttatus*
英文名：Hermit Thrush

2
坦氏孤鸫（雌性）
学　名：*Maydestes townsendi*
英文名：Townsend's Solitaire

3
灰噪鸦（雄性，亚成体）
学　名：*Perisoreus canadensis*
英文名：Grey Jay

Prairie Starling

ICTERUS GUBERNATOR, Aud.

Drawn from Nature by J. J. Audubon F.R.S. F.L.S.

Male 1. Female 2.

Engraved, Printed and Coloured by Robt. Havell 1838

红翅黑鹂

学　名：*Agelaius phoeniceus*

英文名：Red-winged Blackbird

1. 雄性　2. 雌性

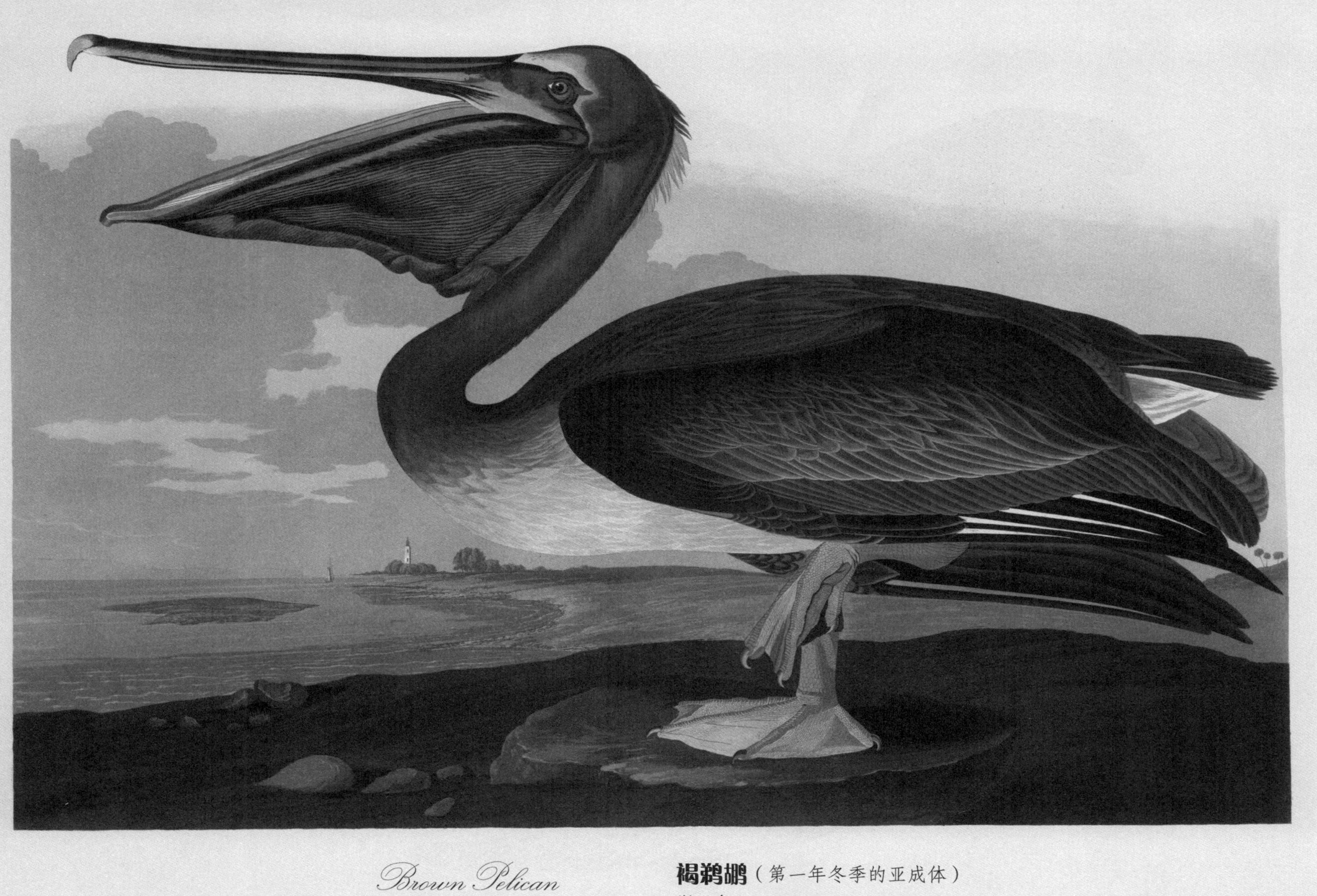

Drawn from Nature by J. J. Audubon F.R.S. F.L.S.

Brown Pelican

PELICANUS FUSCUS

Young first Winter

褐鹈鹕（第一年冬季的亚成体）

学　名：*Pelecanus occidentalis*

英文名：Brown Pelican

Engraved, Printed and Coloured by Robt. Havell, 1838

Rough-legged Falcon

BUTEO LAGOPUS

Drawn from Nature by J. J. Audubon F.R.S. F.L.S.

1. Old Male 2. Young first Winter

Engraved, Printed and Coloured by Robt. Havell, 1838

毛脚𫛭

学　名：*Buteo lagopus*

英文名：Rough-legged Hawk

1. 老年雄性　2. 第一年冬季的亚成体

Thick-legged Partridge

PERDIX NEOXENUS, Aud.

1. Supposed Young Male

Plumed Partridge

PERDIX PLUMIFERA, *Gould*

2. Male 3. Female

Drawn from Nature by J. J. Audubon F.R.S. F.L.S.

Engraved, Printed & Coloured by Robt. Havell, 1838

1

冠齿鹑（假想雄性亚成体）

学　名：*Colinus cristatus*

英文名：Crested Bobwhite

2,3

山翎鹑

学　名：*Oreortyx pictus*

英文名：Mountain Quail

2. 雄性　3. 雌性

Drawn from Nature by J. J. Audubon F.R.S. F.L.S.

Engraved, Printed & Coloured by Robt. Havell, 1838

Lazuli Finch

FRINGILLA AMŒNA, Say

1. Female

Crimson-necked Bull-finch

PYRRHULA FRONTALIS, Bonap

2. Male

Grey-crowned Linnet

LINARIA TEPHROCOTIS, Swains

3. Male

1

白腹蓝彩鹀（雌性）

学　名：*Passerina amoena*

英文名：Lazuli Finch

2

家朱雀（雄性）

学　名：*Carpodacus mexicanus*

英文名：Crimson-necked Bull-finch

3

褐头牛鹂（雄性）

学　名：*Leucosticte tephrocotis*

英文名：Gray-crowned Linnet

Cow-pen Bird

ICTERUS PECORIS, Bonap

4. Young Male

Evening Grosbeak

FRINGILLA VESPERTINA, Cooper

5. Female 6. Young Male

Brown Longspur

PLECTROPHANES TOWNSENDI, Aud.

7. Female

4

粉红腹岭雀（雄性，亚成体）

学　名：*Molothrus ater*

英文名：Cow-pen Bird

5, 6

黄昏锡嘴雀

学　名：*Coccothraustes vespertinus*

英文名：Evening Grosbeak

5. 雌性　6. 雄性，亚成体

7

狐色雀鹀（雌性）

学　名：*Passerella iliaca*

英文名：Fox Sparrow

Columbian Humming Bird

TROCHILUS ANNA, Lesson

1, 2, 3, 4. Male 5. Female and Nest

Plant Hibiscus Virginicus

Drawn from Nature by J. J. Audubon F.R.S. F.L.S.

Engraved, Printed & Coloured by Robt. Havell, 1838

安氏蜂鸟

学　名：*Calypte anna*

英文名：Anna's Hummingbird

1,2,3,4. 雄性　5. 雌性和巢

Californian Vulture

CATHARTES CALIFORNIANUS, Illiger

Drawn from Nature by J. J. Audubon F.R.S. F.L.S.

Old Male

Engraved, Printed and Coloured by Robt. Havell, 1838

加州神鹫（老年雄性）

学　名：*Gymnogyps californianus*

英文名：California Condor

1 *White-legged Oyster-catcher*

HŒMATOPUS BACHMANI, Aud.

1. Male

2 *Slender-billed Oyster-catcher*

HŒMATOPUS TOWNSENDI, Aud.

2. Female

Drawn from Nature by J. J. Audubon F.R.S. F.L.S.

Engraved, Printed and Coloured by Robt. Havell, 1838

北美蛎鹬

学　名：*Haematopus bachmani*

英文名：Black Oystercatcher

1. 雄性　2. 雌性

Townsend's Sandpiper

FRINCA TOWNSENDI, Aud.

Drawn from Nature by J. J. Audubon F.R.S. F.L.S.

Females

Engraved, Printed & Coloured by Robt. Havell, 1838

短嘴鹬（雌性）

学　名：*Aphriza virgata*

英文名：Surfbird

Western Duck

Drawn from Nature by J. J. Audubon F.R.S. F.L.S.

FULIGULA STELLERI, Bonap

Engraved, Printed & Coloured by Robt. Havell, 1838

小绒鸭

学　名：*Polysticta stelleri*

英文名：Steller’s Eider

No. 86.

PLATE CCCCXXX.

2 1

Slender-billed Guillemot

URIA TOWNSENDI, Aud.

1. Male 2. Female

Drawn from Nature by J. J. Audubon F.R.S. F.L.S.

Engraved, Printed & Coloured by Robt. Havell, 1838

斑海雀

学　名：*Brachyramphus marmoratus*

英文名：Marbled Murrelet

1. 雄性　2. 雌性

Drawn from Nature by J. J. Audubon F.R.S. F.L.S.

Engraved, Printed & Coloured by Robt. Havell, 1838

American Flamingo

PHŒNICOPTERUS RUBER, Linn.

Old Male

标注：1. 喙侧面图 2. 上颌外部前视图
3. 上颌内部前视图 4. 下颌内部前视图
5. 下颌内部前视图（包括舌） 6. 舌外视图
7. 舌的外部前视图 8. 舌的内部前视图
9. 脚趾前视图

大红鹳（老年雄性）

学　名：*Phoenicopterus ruber*

英文名：American Flamingo

Drawn from Nature by J. J. Audubon F.R.S. F.L.S.

Engraved, Printed & Coloured by Robt. Havell, 1838

Burrowing Owl　Large-heaede Burrowing Owl　Little night Owl　Columbian Owl　Short-eared Owl

STRIX CUNICULARIA 1. Male

STRIX CALIFORNICA 2. Male

STRIX NOCTUA, Lath. 3. Female

STRIX PASSERINOIDES, Temm. 4, 5. Males

STRIX BRACHYOTUS, Wils. 6. Male

1, 2
穴小鸮（雄性）
学　名：*Athene cunicularia*
英文名：Burrowing Owl

3
纵纹腹小鸮（雌性）
学　名：*Athene noctua*
英文名：Little Owl

4, 5
北美鸺鹠（雄性）
学　名：*Glaucidium californicum*
英文名：Northern Pygmy Owl

6
短耳鸮（雄性）
学　名：*Asio flammeus*
英文名：Short-eared Owl

Drawn from Nature by J. J. Audubon F.R.S. F.L.S.

Engraved, Printed & Coloured by Robt. Havell, 1838

Bullock's Oriole

ICTERUS BULLOCKI, Swains

1. Young Male 2. Old Female

Baltimore Oriole

ICTERUS BALTIMORE, Bonap

3. Old Female

Mexican Goldfinch

CARDUELLIS MEXICANUS, Swains

4. Male 5. Female

Varied Thrush

TURDUS NŒVIUS, Lath.

6. Female

1, 2, 3

橙腹拟鹂

学　名：*Icterus galbula*

英文名：Northern Oriole

1. 雄性，亚成体　2,3. 老年雌性

4, 5

黄脸金翅雀

学　名：*Carduelis yarrellii*

英文名：Yellow-faced Siskin

4. 雄性　5. 雌性

6

杂色鸫（雌性）

学　名：*Ixoreus naevius*

英文名：Varied Thrush

Common Water Thrush

TURDUS AQUATICUS, Wilson

7. Male

7

黄眉灶莺（雄性）

学　名：*Seiurus noveboracensis*

英文名：Northern Waterth rush

Little Tyrant Fly-catcher

TYRANNULA PUSILLA, Swains

1
小纹霸鹟
学　名：*Empidonax minimus*
英文名：Least Flycatcher

Blue Mountain Warbler

SYLVIA MONTANA, Wilson
3. Male

3
蓝山莺（雄性）
学　名：*Sylvia montana*
英文名：Blue Mountain Warble

Short-legged Pewee

MUSCICAPA PHŒBE, Lath.
5. Male

5
绿胁绿霸鹟（雄性）
学　名：*Contopus cooperi*
英文名：Olive-sided Flycatcher

Small-headed Fly-catcher

MUSCICAPA MINUTA, Wils.
2. Male

2
小头莺（雄性）
学　名：*Muscicapa minuta*
英文名：Small-headed Flycatcher

Bartram's Vireo

VIREO BARTRAMI, Swains
4. Male

4
黄喉莺雀（雄性）
学　名：*Vireo flarifrons*
英文名：Yellow-throated Vireo

Rocky Mountain Fly-catcher

TYRANNULA NIGRICANS, Swains
6. Male

6
黑长尾霸鹟（雄性）
学　名：*Sayornis saya*
英文名：Black Phoebe

Columbian Water Ouzel — CINCLUS TOWNSENDI, Aud. — 1. Female

Arctic Water Ouzel — CINCLUS MORTONI, Townsend — 2. Male

Drawn from Nature by J. J. Audubon F.R.S. F.L.S.

Engraved, Printed & Coloured by Robt. Havell, 1838

美洲河乌

学　名：*Cinclus mexicanus*

英文名：American Dipper

1. 雌性　2. 雄性

图版信息

下面每一图版的信息前面是奥杜邦给出的鸟类的英文名和拉丁名，后面则是英国自然博物馆给出的英文名和拉丁名，以及每个图版的原始尺寸（单位：英寸，1 英寸 =2.54 厘米）。图版的编号是其原始编号。

图版编号
奥杜邦给出的英文名
奥杜邦给出的拉丁名
英国自然博物馆给出的英文名
英国自然博物馆给出的拉丁名
图版的原始尺寸（单位：英寸）

PLATE I
Wild turkey (male)
Meleagris gallopavo
Wild turkey
Meleagris gallopavo
38 ⅛ × 25 ½

PLATE II
Yellow-billed cuckoo
Coccyzus carolinensis
Yellow-billed cuckoo
Coccyzus americanus
20 ⅞ × 26 ⅜

PLATE III
Prothonotary warbler
Dacnis protonotarius
Prothonotary warbler
Protonotaria citrea
20 ½ × 12 ½

PLATE IV
Purple finch
Fringilla purpurea
Purple finch
Carpodacus purpureus
20 ⅜ × 12 ½

PLATE V
Bonaparte fly catcher
Muscicapa bonapartii
Canada warbler
Wilsonia canadensis
20 ½ × 12 ½

PLATE VI
Wild turkey (female)
Meleagris gallopavo
Wild turkey
Meleagris gallopavo
25 ⅝ × 38 ¼

PLATE VII
Purple grackle
Quiscalus versicolor
Common grackle
Quiscalus quiscula
26 ⅜ × 20 ¾

PLATE VIII
White throated sparrow
Fringilla pensylvanica
White-throated sparrow
Zonotrichia albicollis
20 ½ × 12 ¾

PLATE IX
Selby's fly catcher
Muscicapa selbii
Hooded warbler
Wilsonia citrina
20 ¾ × 12 ⅞

PLATE X
Brown lark
Anthus aquaticus
Water pipit
Anthus spinoletta
12 ¾ × 20 ½

PLATE 11
The bird of Washington or great American sea eagle
Falco washingtoniensis
Bald eagle
Haliaeetus leucocephalus
38 ¼ × 25 ⅝

PLATE 12
Baltimore oriole
Icterus baltimore
Baltimore oriole
Icterus galbula
26 × 23 ¾

PLATE 13
Snow bird
Fringilla nivalis
Dark-eyed junco
Junco hyemalis
19 ⅜ × 12 ¼

PLATE 14
Prairie warbler
Sylvia discolor
Prairie warbler
Dendroica discolor
19 ⅝ × 12 ½

PLATE XV
Blue yellow back warbler
Sylvia americana
Northern parula
Parula americana
19 ¼ × 12 ¼

PLATE 16
Great footed hawk
Falco peregrinus
Peregrine falcon
Falco peregrinus
25 ⅝ × 38 ¼

PLATE 17
Carolina pigeon or turtle dove
Columba carolinensis
Mourning dove
Zenaida macroura
26 ¾ × 20 ¾

PLATE 18
Bewick's long tailed wren
Troglodytes bewickii
Bewick's wren
Thryomanes bewickii
19 ⅝ × 12 ⅛

PLATE 19
Louisiana water thrush
Turdus aquaticus
Louisiana waterthrush
Seiurus motacilla
19 ¾ × 12 ½

PLATE 20
Blue winged yellow warbler
Dacnis solitaria
Blue-winged warbler
Vermivora pinus
19 ½ × 12 ¼

PLATE 21
The mocking bird
Turdus polyglottus
Northern mockingbird
Mimus polyglottos
33 ¼ × 23 ⅝

PLATE 22
Purple martin
Hirundo purpurea
Purple martin
Progne subis
25 ½ × 20 ½

PLATE 23
Maryland yellow throat
Sylvia trichas
Common yellowthroat
Geothlypis trichas
19 ⅜ × 12 ⅛

PLATE 24
Roscoe's yellow throat
Sylvia rosco
Common yellowthroat
Geothlypis trichas
19 ⅝ × 12 ¼

PLATE 25
Song sparrow
Fringilla melodia
Song sparrow
Melospiza melodia
19 ⅜ × 12 ⅛

PLATE 26
Carolina parrot
Psitacus carolinensis
Carolina parakeet
Conuropsis carolinensis
33 ¼ × 24

PLATE 27
Red headed woodpecker
Picus erythrocephalus
Red-headed woodpecker
Melanerpes erythrocephalus
25 ⅜ × 21 ¾

PLATE 28
Solitary flycatcher
Vireo solitarius
Blue-headed vireo
Vireo solitarius
19 ½ × 12 ¼

PLATE 29
Towee bunting
Fringilla erythropthalma
Eastern towhee
Pipilo erythrophthalmus
19 ⅝ × 12 ⅜

PLATE 30
Vigors vireo
Vireo vigorsii
Pine warbler
Dendroica pinus
19 ½ × 12 ¼

PLATE 31
White-headed eagle
Falco leucocephalus
Bald eagle
Haliaeetus leucocephalus
25 ⅝ × 38 ¼

PLATE 32
Black-billed cuckoo
Coccyzus erythrophthalmus
Black-billed cuckoo
Coccyzus erythropthalmus
18 ¾ × 26 ⅜

PLATE 33
Yellow bird or American goldfinch
Carduelis americana
American goldfinch
Carduelis tristis
19 ⅜ × 12 ¼

PLATE 34
Worm-eating warbler
Dacnis vermivora
Worm-eating warbler
Helmitheros vermivorum
19 ⅝ × 12 ¼

PLATE 35
Children's warbler
Silvia childreni
Yellow warbler
Dendroica petechia
19 ½ × 12 ¼

PLATE 36
Stanley hawk
Astur stanleii
Cooper's hawk
Accipiter cooperii
38 ¼ × 25 ⅝

PLATE 37
Gold-winged woodpecker
Picus auratus
Northern flicker
Colaptes auratus
25 ⅞ × 20 ¾

PLATE 38
Kentucky warbler
Sylvia formosa
Kentucky warbler
Oporornis formosus
19 ½ × 12 ¼

PLATE 39
Crested titmouse
Parus bicolor
Tufted titmouse
Parus bicolor
19 ½ × 12 ¼

PLATE 40
American redstart
Muscicapa ruticilla
American redstart
Setophaga ruticilla
19 ⅜ × 12 ⅛

PLATE 41
Ruffed grous
Tetrao umbellus
Ruffed grouse
Bonasa umbellus
25 ⅝ × 38 ¼

PLATE 42
Orchard oriole
Icterus spurius
Orchard oriole
Icterus spurius
26 × 20 ⅞

PLATE 43
Cedar bird
Bombycilla carolinensis
Cedar waxwing
Bombycilla cedrorum
19 ⅝ × 12 ¼

PLATE 44
Summer red bird
Tanagra æstiva
Summer tanager
Piranga rubra
19 ⅝ × 12 ¼

PLATE 45
Traill's fly-catcher
Muscicapa trailli
Willow flycatcher
Empidonax traillii
19 ½ × 12 ¼

PLATE 46
Barred owl
Strix nebulosa
Barred owl
Strix varia
38 ¼ × 25 ⅝

PLATE 47
Ruby-throated humming bird
Trochilus colubris
Ruby-throated hummingbird
Archilochus colubris
25 ⅞ × 20 ¾

PLATE 48
Cerulean warbler
Sylvia azurea
Cerulean warbler
Dendroica cerulea
19 ¾ × 12 ¼

PLATE 49
Blue-green warbler
Sylvia rara
Cerulean warbler
Dendroica cerulea
19 ⅝ × 12 ¼

PLATE 50
Swainson's warbler
Sylvicola swainsonia
Magnolia warbler
Dendroica magnolia
19 ½ × 12 ¼

PLATE 51
Red tailed hawk
Falco borealis
Red-tailed hawk
Buteo jamaicensis
38 ⅛ × 25 ½

PLATE 52
Chuck will's widow
Caprimulgus carolinensis
Chuck-will's widow
Caprimulgus carolinensis
26 × 20 ⅝

PLATE 53
Painted bunting
Fringilla ciris
Painted bunting
Passerina ciris
19 ⅜ × 12 ⅛

PLATE 54
Rice bunting
Icterus agripennis
Bobolink
Dolichonyx oryzivorus
19 ½ × 12 ¼

PLATE 55
Cuvier's wren
Regulus cuvieri
Cuvier's kinglet
Regulus cuvieri
19 ½ × 12 ⅛

PLATE 56
Red-shouldered hawk
Falco lineatus
Red-shouldered hawk
Buteo lineatus
38 ⅛ × 25 ½

PLATE 57
Loggerhead shrike
Lanius carolinensis
Loggerhead shrike
Lanius ludovicianus
26 × 20 ⅝

PLATE 58
Hermit thrush
Turdus solitarius
Hermit thrush
Catharus guttatus
19 ⅜ × 12 ⅛

PLATE 59
Chesnut sided warbler
Sylvia Icterocephala
Chestnut-sided warbler
Dendroica pensylvanica
19 ½ × 12 ¼

PLATE 60
Carbonated warbler
Sylvia carbonata
Carbonated warbler
Sylvia carbonata
19 ½ × 12 ¼

PLATE 61
Great horned-owl
Strix virginiana
Great horned owl
Bubo virginianus
38 ⅛ × 25 ½

PLATE 62
Pafsenger pigeon
Columba migratoria
Passenger pigeon
Ectopistes migratorius
26 × 20 ⅞

PLATE 63
White eyed flycatcher
Vireo noveboracensis
White-eyed vireo
Vireo griseus
19 ⅜ × 12 ¼

PLATE 64
Swamp sparrow
Spiza palustris
Swamp sparrow
Melospiza georgiana
19 ⅜ × 12 ¼

PLATE 65
Rathbone's warbler
Sylvia rathboni
Yellow warbler
Dendroica petechia
19 ¼ × 12 ⅛

PLATE 66
Ivory-billed woodpecker
Picus principalis
Ivory-billed woodpecker
Campephilus principalis
38 ¼ × 25 ⅝

PLATE 67
Red-winged starling
Icterus phœniceus
Red-winged blackbird
Agelaius phoeniceus
25 ⅞ × 20 ¾

PLATE 68
Republican cliff swallow
Hirundo fulva
Cliff swallow
Petrochelidon pyrrhonota
19 ½ × 12 ¼

PLATE 69
Bay breasted warbler
Sylvia castanea
Bay-breasted warbler
Dendroica castanea
19 ½ × 12 ¼

PLATE 70
Henslow's bunting
Ammodramus henslowii
Henslow's sparrow
Ammodramus henslowii
19 ½ × 12 ¼

PLATE 71
Winter hawk
Circus hyemalis
Red-shouldered hawk
Buteo lineatus
25 ½ × 38 ⅛

PLATE 72
Swallow-tailed hawk
Falco furcatus
Swallow-tailed kite
Elanoides forficatus
20 ¾ × 27 ⅛

PLATE 73
Wood thrush
Turdus mustelinus
Wood thrush
Hylocichla mustelina
19 ½ × 12 ¼

PLATE 74
Indigo-bird
Fringilla cyanea
Indigo bunting
Passerina cyanea
19 ⅝ × 12 ¼

PLATE 75
Le petit caporal
Falco temerarius
Merlin
Falco columbarius
19 ½ × 12 ¼

PLATE 76
Virginian partridge
Perdix virginiana
Northern bobwhite
(also red-shouldered hawk)
Colinus virginianus
(also *Buteo lineatus*)
25 ⅝ × 38 ¼

PLATE 77
Belted kingfisher
Alcedo alcyon
Belted kingfisher
Megaceryle alcyon
25 ⅞ × 20 ¾

PLATE 78
Great carolina wren
Troglodytes ludovicianus
Carolina wren
Thryothorus ludovicianus
19 ⅜ × 12 ⅛

PLATE 79
Tyrant flycatcher
Muscicapa tyrannus
Eastern kingbird
Tyrannus tyrannus
19 ½ × 12 ¼

PLATE 80
Anthus hypogaeus
Phlox subulata
Water pipit
Anthus spinoletta
12 ⅛ × 19 ⅜

PLATE 81
Fish hawk
Falco haliætus
Osprey
Pandion haliaetus
38 ¼ × 25 ⅝

PLATE 82
Whip-poor-will
Caprimulgus vociferus
Whip-poor-will
Caprimulgus vociferus
25 ¾ × 20 ¾

PLATE 83
House wren
Troglodytes ædon
House wren
Troglodytes aedon
19 ½ × 12 ¼

PLATE 84
Blue grey flycatcher
Sylvia cœrula
Blue-grey gnatcatcher
Polioptila caerulea
19 ½ × 12 ¼

PLATE 85
Yellow throat warbler
Sylvia pensilis
Yellow-throated warbler
Dendroica dominica
19 ⅜ × 12 ¼

PLATE 86
Black warrior
Falco harlani
Red-tailed hawk
Buteo jamaicensis
38 ⅛ × 25 ½

PLATE 87
Florida jay
Garrulus floridanus
Florida scrub jay
Aphelocoma coerulescens
25 ¾ × 20 ½

PLATE 88
Autumnal warbler
Sylvia autumnalis
Bay-breasted warbler
Dendroica castanea
19 ⅜ × 12 ¼

PLATE 89
Nashville warbler
Sylvia rubricapilla
Nashville warbler
Vermivora ruficapilla
19 ⅜ × 12 ¼

PLATE 90
Black and white creeper
Sylvia varia
Black-and-white warbler
Mniotilta varia
19 ½ × 12 ¼

PLATE 91
Broad-winged hawk
Falco pennsylvanicus
Broad-winged hawk
Buteo platypterus
38 ⅛ × 25 ½

PLATE 92
Pigeon hawk
Falco columbarius
Merlin
Falco columbarius
25 ⅜ × 20 ⅜

PLATE 93
Sea-side finch
Fringilla maritima
Seaside sparrow
Ammodramus maritimus
19 ¼ × 12 ¼

PLATE 94
Bay-winged bunting
Fringilla graminea
Vesper sparrow
Pooecetes gramineus
19 ½ × 12 ¼

PLATE 95
Blue-eyed yellow warbler
Sylvia æstiva
Yellow warbler
Dendroica petechia
19 ⅜ × 12 ¼

PLATE 96
Columbia jay
Garrulus ultramarinus
Black-throated magpie-jay
Calocitta colliei
38 ⅛ × 25 ½

PLATE 97
Mottled owl
Strix asio
Eastern screech owl
Otus asio
25 ⅜ × 20 ¾

PLATE 98
Marsh wren
Troglodytes palustris
Marsh wren
Cistothorus palustris
19 ½ × 12 ¼

PLATE 99
Cow bunting
Icterus pecoris
Brown-headed cowbird
Molothrus ater
12 ¼ × 19 ⅝

PLATE 100
Green-blue, or white, bellied swallow
Hirundo bicolor
Tree swallow
Tachycineta bicolor
19 ⅝ × 12 ¼

PLATE CI
Raven
Corvus corax
Common raven
Corvus corax
38 ⅛ × 25 ⅜

PLATE CII
Blue jay
Corvus cristatus
Blue jay
Cyanocitta cristata
25 ½ × 20 ½

PLATE CIII
Canada warbler
Sylvia pardalina
Canada warbler
Wilsonia canadensis
19 ¼ × 12 ⅛

PLATE CIV
Chipping sparrow
Fringilla socialis
Chipping sparrow
Spizella passerina
19 ½ × 12 ¼

PLATE CV
Red-breasted nuthatch
Sitta canadensis
Red-breasted nuthatch
Sitta canadensis
19 ⅝ × 12 ¼

PLATE CVI
Black vulture or carrion crow
Cathartes atratus
Black vulture
Coragyps atratus
25 ⅝ × 38 ¼

PLATE CVII
Canada jay
Corvus canadensis
Grey jay
Perisoreus Canadensis
26 × 20 ¾

PLATE CVIII
Fox-coloured sparrow
Fringilla iliaca
Fox sparrow
Passerella iliaca
12 ¼ × 19 ½

PLATE CIX
Savannah finch
Fringilla savanna
Savannah sparrow
Passerculus sandwichensis
19 ⅝ × 12 ⅜

PLATE CX
Hooded warbler
Sylvia mitrata
Hooded warbler
Wilsonia citrina
19 ⅜ × 12 ¼

PLATE CXI
Pileated woodpecker
Picus pileatus
Pileated woodpecker
Dryocopus pileatus
38 ¼ × 25 ⅝

PLATE CXII
Downy woodpecker
Picus pubescens
Downy woodpecker
Picoides pubescens
26 × 20 ⅞

PLATE CXIII
Blue-bird
Sylvia sialis
Eastern bluebird
Sialia sialis
19 ½ × 12 ⅜

PLATE CXIV
White-crowned sparrow
Fringilla leucophrys
White-crowned sparrow
Zonotrichia leucophrys
19 ½ × 12 ¼

PLATE CXV
Wood pewee
Muscicapa virens
Eastern wood pewee
Contopus virens
19 ½ × 12 ¼

PLATE CXVI
Ferruginous thrush
Turdus rufus
Brown thrasher
Toxostoma rufum
38 ¼ × 25 ⅝

PLATE CXVII
Mississippi kite
Falco plumbeus
Mississippi kite
Ictinia mississippiensis
25 ⅞ × 20 ⅞

PLATE CXVIII
Warbling flycatcher
Muscicapa gilva
Warbling vireo
Vireo gilvus
19 ⅜ × 12 ¼

PLATE CXIX
Yellow-throated vireo
Vireo flavifrons
Yellow-throated vireo
Vireo flavifrons
19 ½ × 12 ⅛

PLATE CXX
Pewit flycatcher
Muscicapa fusca
Eastern phoebe
Sayornis phoebe
19 ½ × 12 ¼

PLATE CXXI
Snowy owl
Strix nyctea
Snowy owl
Bubo scandiacus
38 ¼ × 25 ⅝

PLATE CXXII
Blue grosbeak
Fringilla corulea
Blue grosbeak
Passerina caerulea
26 × 20 ⅞

PLATE CXXIII
Black and yellow warbler
Sylvia maculosa
Magnolia warbler
Dendroica magnolia
19 ⅝ × 12 ¼

PLATE CXXIV
Green black-capt flycatcher
Muscicapa pusilla
Wilson's warbler
Wilsonia pusilla
19 ⅝ × 12 ⅜

PLATE CXXV
Brown-headed nuthatch
Sitta pusilla
Brown-headed nuthatch
Sitta pusilla
19 ⅝ × 12 ⅜

PLATE CXXVI
White-headed eagle
Falco leucocephalus
Bald eagle
Haliaeetus leucocephalus
38 ¼ × 25 ⅝

PLATE CXXVII
Rose-breasted grosbeak
Fringilla ludoviciana
Rose-breasted grosbeak
Pheucticus ludovicianus
25 ¾ × 20 ⅝

PLATE CXXVIII
Cat bird
Turdus felivox
Gray catbird
Dumetella carolinensis
19 ⅝ × 12 ⅜

PLATE CXXIX
Great crested flycatcher
Muscicapa crinita
Great crested flycatcher
Myiarchus crinitus
19 ½ × 12 ¼

PLATE CXXX
Yellow-winged sparrow
Fringilla passerina
Grasshopper sparrow
Ammodramus savannarum
19 ½ × 12 ¼

PLATE CXXXI
American robin
Turdus migratorius
American robin
Turdus migratorius
38 ⅛ × 25 ⅝

PLATE CXXXII
Three-toed woodpecker
Picus tridactylus
Black-backed woodpecker
Picoides arcticus
26 × 20 ¾

PLATE CXXXIII
Black-poll warbler
Sylvia striata
Blackpoll warbler
Dendroica striata
19 ⅝ × 12 ⅜

PLATE CXXXIV
Hemlock warbler
Sylvia parus
Blackburnian warbler
Dendroica fusca
19 ½ × 12 ⅜

PLATE CXXXV
Blackburnian warbler
Sylvia blackburnia
Blackburnian warbler
Dendroica fusca
19 ¾ × 12 ⅜

PLATE CXXXVI
Meadow lark
Sturnus ludovicianus
Eastern meadowlark
Sturnella magna
38 ¼ × 25 ⅝

PLATE CXXXVII
Yellow-breasted chat
Icteria viridis
Yellow-breasted chat
Icteria virens
25 ¾ × 20 ⅝

PLATE CXXXVIII
Connecticut warbler
Sylvia agilis
Connecticut warbler
Oporornis agilis
19 ½ × 12 ¼

PLATE CXXXIX
Field sparrow
Fringilla pusilla
Field sparrow
Spizella pusilla
19 ½ × 12 ¼

PLATE CXL
Pine creeping warbler
Sylvia pinus
Pine warbler
Dendroica pinus
19 ½ × 12 ⅜

PLATE CXLI
Species 1
Goshawk
Falco palumbarius
Northern goshawk
Accipiter gentilis

Species 2
Stanley hawk
Falco stanleii
Cooper's hawk
Accipiter cooperii
38 ¼ × 25 ⅝

PLATE CXLII
American sparrow hawk
Falco sparverius
American kestrel
Falco sparverius
26 × 20 ⅝

PLATE CXLIII
Golden-crowned thrush
Turdus aurocapillus
Ovenbird
Seiurus aurocapilla
19 ½ × 12 ¼

PLATE CXLIV
Small green crested flycatcher
Muscicapa acadica
Acadian flycatcher
Empidonax virescens
19 ⅝ × 12 ⅜

PLATE CXLV
Yellow red-poll warbler
Sylvia petechia
Palm warbler
Dendroica palmarum
19 ½ × 12 ⅜

PLATE CXLVI
Fish crow
Corvus ossifragus
Fish crow
Corvus ossifragus
38 ¼ × 25 ⅝

PLATE CXLVII
Night hawk
Caprimulgus virginianus
Common nighthawk
Chordeiles minor
26 × 20 ⅝

PLATE CXLVIII
Pine swamp warbler
Sylvia sphagnosa
Black-throated blue warbler
Dendroica caerulescens
19 ⅝ × 12 ¼

PLATE CXLIX
Sharp-tailed finch
Fringilla caudacuta
Saltmarsh sharp-tailed sparrow
Ammodramus caudacutus
19 ⅝ × 12 ¼

PLATE CL
Red-eyed vireo
Vireo olivaceus
Red-eyed vireo
Vireo olivaceus
19 ½ × 12 ¼

PLATE CLI
Turkey buzzard
Cathartes atratus
Turkey-vulture
Cathartes aura
25 ⅝ × 38 ¼

PLATE CLII
White-breasted black-capped nuthatch
Sitta carolinensis
White-breasted nuthatch
Sitta carolinensis
26 × 20 ¾

PLATE CLIII
Yellow-crown warbler
Sylvia coronata
Yellow-rumped warbler
Dendroica coronata
19 ⅝ × 12 ⅜

PLATE CLIV
Tennessee warbler
Sylvia peregrina
Tennessee warbler
Vermivora peregrina
19 ⅝ × 12 ½

PLATE CLV
Black-throated blue warbler
Sylvia canadensis
Black-throated blue warbler
Dendroica caerulescens
19 ⅝ × 12 ⅜

PLATE CLVI
American crow
Corvus americanus
American crow
Corvus brachyrhynchos
38 ¼ × 25 ⅝

PLATE CLVII
Rusty grakle
Quiscalus ferrugineus
Rusty blackbird
Euphagus carolinus
25 ⅞ × 20 ⅞

PLATE CLVIII
American swift
Cypselus pelasgius
Chimney swift
Chaetura pelagica
19 ½ × 12 ⅜

PLATE CLIX
Cardinal grosbeak
Fringilla cardinalis
Northern cardinal
Cardinalis cardinalis
19 ⅜ × 12 ¼

PLATE CLX
Black-capped titmouse
Parus atricapillus
Carolina chickadee
Parus carolinensis
19 ½ × 12 ¼

PLATE CLXI
Brasilian caracara eagle
Polyborus vulgaris
Crested caracara
Caracara cheriway
38 ¼ × 25 ⅝

PLATE CLXII
Zenaida dove
Columba zenaida
Zenaida dove
Zenaida aurita
25 ⅞ × 20 ¾

PLATE CLXIII
Palm warbler
Sylvia palmarum
Palm warbler
Dendroica palmarum
19 ½ × 12 ¼

PLATE CLXIV
Tawny thrush
Turdus wilsonii
Veery
Catharus fuscescens
19 ½ × 16 ⅛

PLATE CLXV
Bachmans finch
Fringilla bachmani
Bachman's sparrow
Aimophila aestivalis
19 ⅝ × 12 ⅜

PLATE CLXVI
Rough-legged falcon
Flaco lagopus
Rough-legged buzzard
Buteo lagopus
38 ¼ × 25 ⅝

PLATE CLXVII
Key-west dove
Columba montana
Key West quail-dove
Geotrygon chrysia
20 ⅞ × 25 ⅞

PLATE CLXVIII
Forked-tailed flycatcher
Muscicapa savana
Fork-tailed flycatcher
Tyrannus savana
19 ⅝ × 12 ¼

PLATE CLXIX
Mangrove cuckoo
Coccyzus seniculus
Mangrove cuckoo
Coccyzus minor
19 ½ × 12 ⅜

PLATE CLXX
Gray tyrant
Tyrannus grisens
Grey kingbird
Tyrannus dominicensis
19 ¾ × 12 ¼

PLATE CLXXI
Barn owl
Strix flammea
Barn owl
Tyto alba
38 ¼ × 25 ⅝

PLATE CLXXII
Blue-headed pigeon
Columba cyanocephala
Blue-headed quail-dove
Starnoenas cyanocephala
20 ⅞ × 26

PLATE CLXXIII
Barn swallow
Hirundo americana
Barn swallow
Hirundo rustica
19 ½ × 12 ¼

PLATE CLXXIV
Olive sided flycatcher
Muscicapa inornata
Olive-sided flycatcher
Contopus cooperi
19 ⅝ × 12 ¼

PLATE CLXXV
Nuttalls lesser-marsh wren
Troglodites brevirostris
Sedge wren
Cistothorus platensis
19 ½ × 12 ¼

PLATE CLXXVI
Spotted grous
Tetrao canadensis
Spruce grouse
Canachites canadensis
25 ½ × 38 ⅜

PLATE CLXXVII
White-crowned pigeon
Columba leucocephala
White-crowned pigeon
Columba leucocephala
25 ⅝ × 20 ¾

PLATE CLXXVIII
Orange-crowned warbler
Sylvia celata
Orange-crowned warbler
Vermivora celata
19 ⅝ × 12 ⅜

PLATE CLXXIX
Wood wren
Troglodytes americana
House wren
Troglodytes aedon
19 ½ × 12 ⅜

PLATE CLXXX
Pine finch
Fringilla pinus
Pine siskin
Carduelis pinus
19 ½ × 12 ¼

PLATE CLXXXI
Golden eagle
Aquila chrysaetos
Golden eagle
Aquila chrysaetos
37 ¼ × 25 ⅝

PLATE CLXXXII
Ground dove
Columba passerina
Common ground dove
Columbina passerina
25 ⅞ × 20 ⅞

PLATE CLXXXIII
Golden crested-wren
Regulus cristatus
Golden-crowned kinglet
Regulus satrapa
19 ½ × 12 ¼

PLATE CLXXXIV
Mangrove humming bird
Trochilus mango
Black-throated mango
Anthracothorax nigricollis
18 ⅜ × 13 ¾

PLATE CLXXXV
Bachman's warbler
Sylvia bachmanii
Bachman's warbler
Vermivora bachmanii
20 ½ × 14 ¾

PLATE CLXXXVI
Pinnated grous
Tetrao cupido
Greater prairie chicken
Tympanuchus cupido
25 ½ × 37 ¼

PLATE CLXXXVII
Boat-tailed grackle
Quiscalus major
Boat-tailed grackle
Quiscalus major
26 × 20 ⅞

PLATE CLXXXVIII
Tree sparrow
Fringilla canadensis
American tree sparrow
Spizella arborea
19 ½ × 12 ¼

PLATE CLXXXIX
Snow bunting
Emberiza nivalis
Snow bunting
Plectrophenax nivalis
19 ½ × 12 ⅜

PLATE CXC
Yellow bellied woodpecker
Picus varius
Yellow-bellied sapsucker
Sphyrapicus varius
19 ⅝ × 12 ⅜

PLATE CXCI
Willow grous or large ptarmigan
Tetrao saliceti
Willow grouse
Lagopus lagopus
25 ⅝ × 38 ¼

PLATE CXCII
Great American shrike or butcher bird
Lanius septentuonalis
Great grey shrike
Lanius excubitor
26 × 20 ¾

PLATE CXCIII
Lincoln finch
Fringilla lincolnii
Lincoln's sparrow
Melospiza lincolnii
19 ½ × 12 ½

PLATE CXCIV
Canadian titmouse
Parus hudsonicus
Boreal chickadee
Parus hudsonicus
19 ½ × 12 ⅜

PLATE CXCV
Ruby crowned wren
Regulus calendula
Ruby-crowned kinglet
Regulus calendula
19 ½ × 12 ½

PLATE CXCVI
Labrador falcon
Falco labradora
Gyrfalcon
Falco rusticolus
38 ¼ × 25 ⅝

PLATE CXCVII
American crossbill
Loxia curvirostra
Red crossbill
Loxia curvirostra
25 ⅞ × 20 ¾

PLATE CXCVIII
Brown headed worm eating warbler
Sylvia swainsonii
Swainson's warbler
Limnothlypis swainsonii
19 ⅝ × 12 ¼

PLATE CXCIX
Little owl
Strix acadica
Northern saw-whet owl
Aegolius acadicus
19 ⅝ × 12 ¼

PLATE CC
Shore lark
Alauda alpestris
Horned lark
Eremophila alpestris
12 ¼ × 19 ½

PLATE CCI
Canada goose
Anser canadensis
Canada goose
Branta canadensis
38 × 25 ½

PLATE CCII
Red-throated diver
Colymbus septentrionalis
Red-throated diver
Gavia stellata
20 ⅞ × 28 ⅜

PLATE CCIII
Fresh water marsh hen
Rallus elegans
King rail
Rallus elegans
12 ¾ × 19 ½

PLATE CCIV
Salt water marsh hen
Rallus crepitans
Clapper rail
Rallus longirostris
19 ½ × 14 ⅞

PLATE CCV
Virginia rail
Rallus virginianus
Virginia rail
Rallus limicola
14 ¾ × 20 ⅜

PLATE CCVI
Summer or wood duck
Anas sponsa
Wood duck
Aix sponsa
30 ⅛ × 25 ½

PLATE CCVII
Booby gannet
Sula fusca
Brown booby
Sula leucogaster
25 ¾ × 20 ½

PLATE CCVIII
Esquimaux curlew
Numenius borealis
Eskimo curlew
Numenius borealis
10 ¼ × 19 ½

PLATE CCIX
Wilson's plover
Charadrius wilsonius
Wilson's plover
Charadrius wilsonia
12 ⅛ × 19 ⅜

PLATE CCX
Least bittern
Ardea exilis
Least bittern
Ixobrychus exilis
12 ¼ × 19 ½

PLATE CCXI
Great blue heron
Ardea herodias
Great blue heron
Ardea herodias
38 ⅛ × 25 ½

PLATE CCXII
Common gull
Larus canus
Ring-billed gull
Larus delawarensis
20 ⅞ × 26

PLATE CCXIII
Puffin
Mormon arcticus
Atlantic puffin
Fratercula arctica
12 ¼ × 19 ½

PLATE CCXIV
Razor bill
Alca torda
Razorbill
Alca torda
12 ¼ × 19 ½

PLATE CCXV
Hyperborean phalarope
Phalaropus hyperboreus
Red-necked phalarope
Phalaropus lobatus
12 ¼ × 19 ½

PLATE CCXVI
Wood ibiss
Tantalus loculator
Wood stork
Mycteria americana
25 ½ × 38 ⅛

PLATE CCXVII
Louisiana heron
Ardea ludoviciana
Tricolored heron
Egretta tricolor
20 ¾ × 25 ⅞

PLATE CCXVIII
Foolish guillemot
Uria troile
Guillemot
Uria aalge
12 ¼ × 19 ½

PLATE CCXIX
Black guillemot
Uria grylle
Black guillemot
Cepphus grylle
17 ½ × 20 ⅜

PLATE CCXX
Piping plover
Charadrius melodus
Piping plover
Charadrius melodus
12 ¼ × 19 ½

PLATE CCXXI
Mallard duck
Anas boschas
Mallard
Anas platyrhynchos
25 ½ × 38 ⅛

PLATE CCXXII
White ibis
Ibis alba
White ibis
Eudocimus albus
20 ¾ × 25 ¾

PLATE CCXXIII
Pied oyster-catcher
Hæmatopus ostralegus
American oystercatcher
Haematopus palliatus
12 ¼ × 19 ⅜

PLATE CCXXIV
Kittiwake gull
Larus tridactylus
Black-legged kittiwake
Rissa tridactyla
12 ¼ × 19 ⅜

PLATE CCXXV
Kildeer plover
Charadrius vociferus
Killdeer
Charadrius vociferus
12 ¼ × 19 ⅜

PLATE CCXXVI
Hooping crane
Grus americana
Whooping crane
Grus americana
38 ⅛ × 25 ½

PLATE CCXXVII
Pin tailed duck
Anas acuta
Northern pintail
Anas acuta
20 ¾ × 25 ¾

PLATE CCXXVIII
American green winged teal
Anas carolinensis
Green-winged teal
Anas carolinensis
12 ¼ × 19 ⅜

PLATE CCXXIX
Scaup duck
Fuligula marila
Greater scaup
Aythya marila
12 ¼ × 19 ½

PLATE CCXXX
Ruddy plover
Tringa arenaria
Sanderling
Calidris alba
12 ¼ × 19 ⅜

PLATE CCXXXI
Long-billed curlew
Numenius longirostris
Long-billed curlew
Numenius americanus
25 ½ × 38 ⅛

PLATE CCXXXII
Hooded merganser
Mergus cucullatus
Hooded merganser
Lophodytes cucullatus
20 ⅞ × 26

PLATE CCXXXIII
Sora or rail
Rallus carolinus
Sora rail
Porzana carolina
12 ¼ × 19 ½

PLATE CCXXXIV
Tufted duck
Fuligula rufitorques
Ring-necked duck
Aythya collaris
13 ¾ × 17 ⅜

PLATE CCXXXV
Sooty tern
Sterna fuliginosa
Sooty tern
Sterna fuscata
12 ¼ × 19 ⅜

PLATE CCXXXVI
Night heron or qua bird
Ardea nycticorax
Black-crowned night heron
Nycticorax nycticorax
25 ½ × 38 ⅛

PLATE CCXXXVII
Great esquimaux curlew
Numenius hudsonicus
Whimbrel
Numenius phaeopus
20 ⅝ × 25 ¾

PLATE CCXXXVIII
Great marbled godwit
Limosa fedoa
Marbled godwit
Limosa fedoa
13 ¼ × 20 ⅞

PLATE CCXXXIX
American coot
Fulica americana
American coot
Fulica americana
12 ¼ × 19 ¼

PLATE CCXL
Roseate tern
Sterna dougallii
Roseate tern
Sterna dougallii
19 ½ × 12 ¼

PLATE CCXLI
Black backed gull
Larus marinus
Great black-backed gull
Larus marinus
38 ⅛ × 25 ½

PLATE CCXLII
Snowy heron or white egret
Ardea candidissima
Snowy egret
Egretta thula
25 ¾ × 20 ½

PLATE CCXLIII
American snipe
Scolopax wilsonii
Common snipe
Gallinago gallinago
12 ¼ × 19 ⅜

PLATE CCXLIV
Common gallinule
Gallinula chloropus
Common moorhen
Gallinula chloropus
12 ¼ × 19 ⅜

PLATE CCXLV
[No common name given]
Uria brunnichii
Brunnich's guillemot
Uria lomvia
12 ¼ × 21 ½

PLATE CCXLVI
Eider duck
Fuligula mollissima
Common eider
Somateria mollissima
25 ½ × 38 ⅛

PLATE CCXLVII
Velvet duck
Fuligula fusca
Velvet scoter
Melanitta fusca
20 ⅝ × 29 ⅞

PLATE CCXLVIII
American pied-bill dobchick
Podiceps carolinensis
Pied-billed grebe
Podilymbus podiceps
14 ⅝ × 22 ¼

PLATE CCXLIX
Tufted auk
Mormon cirrhatus
Tufted puffin
Fratercula cirrhata
14 ¼ × 19 ½

PLATE CCL
Arctic tern
Sterna arctica
Arctic tern
Sterna paradisaea
19 ⅜ × 12 ¼

PLATE CCLI
Brown pelican
Pelecanus fuscus
Brown pelican
Pelecanus occidentalis
38 ⅛ × 25 ½

PLATE CCLII
Florida cormorant
Carbo floridanus
Double-crested cormorant
Phalacrocorax auritus
19 ¾ × 26 ⅛

PLATE CCLIII
Jager
Lestris pomarina
Pomarine skua
Stercorarius pomarinus
15 ¾ × 21 ¼

PLATE CCLIV
Wilson's phalarope
Phalaropus wilsonii
Wilson's phalarope
Phalaropus tricolor
15 ¾ × 21 ¼

PLATE CCLV
Red phalarope
Phalaropus platyrhynchus
Grey phalarope
Phalaropus fulicarius
15 ¾ × 22 ½

PLATE CCLVI
Purple heron
Ardea rufescens
Reddish egret
Egretta rufescens
25 ½ × 38 ¼

PLATE CCLVII
Double-crested cormorant
Phalacrocorax dilophus
Double-crested cormorant
Phalacrocorax auritus
30 ¼ × 21 ⅝

PLATE CCLVIII
Hudsonian godwit
Limosa hudsonica
Hudsonian godwit
Limosa haemastica
14 ¾ × 20 ⅜

PLATE CCLIX
Horned grebe
Podiceps cornutus
Horned grebe
Podiceps auritus
14 ¾ × 20 ⅜

PLATE CCLX
Fork-tail petrel
Thalassidroma leachii
Leach's storm petrel
Oceanodroma leucorhoa
12 ¼ × 19 ⅜

PLATE CCLXI
Hooping crane
Grus americana
Sandhill crane
Grus canadensis
38 ⅛ × 25 ½

PLATE CCLXII
Tropic bird
Phaeton æthereus
White-tailed tropicbird
Phaethon lepturus
20 ⅝ × 29 ⅞

PLATE CCLXIII
Pigmy curlew
Tringa subarquata
Curlew sandpiper
Calidris ferruginea
12 ⅛ × 19 ¼

PLATE CCLXIV
Fulmar petrel
Procellaria glacialis
Northern fulmar
Fulmarus glacialis
12 ¼ × 19 ⅜

PLATE CCLXV
Buff breasted sandpiper
Tringa rufescens
Buff-breasted sandpiper
Tryngites subruficollis
12 ¼ × 19 ⅜

PLATE CCLXVI
Common cormorant
Phalacrocorax carbo
Great cormorant
Phalacrocorax carbo
25 × 38 ⅛

PLATE CCLXVII
Arctic yager
Lestris parasitica
Long-tailed skua
Stercorarius longicaudus
30 × 21 ⅝

PLATE CCLXVIII
American woodcock
Scolopax minor
American woodcock
Scolopax minor
14 ⅝ × 20 ⅜

PLATE CCLXIX
Greenshank
Totanus glottis
Common greenshank
Tringa nebularia
14 ¾ × 20 ⅝

PLATE CCLXX
Stormy petrel
Thalassidroma wilsonii
Wilson's storm petrel
Oceanites oceanicus
12 ⅜ × 19 ½

PLATE CCLXXI
Frigate pelican
Tachypetes aquilus
Magnificent frigatebird
Fregata magnificens
38 ⅛ × 25 ½

PLATE CCLXXII
Richardson's jager
Lestris richardsonii
Arctic skua
Stercorarius parasiticus
20 ⅝ × 25 ¾

PLATE CCLXXIII
Cayenne tern
Sterna cayana
Royal tern
Sterna maxima
14 ⅞ × 20 ½

PLATE CCLXXIV
Semipalmated shipe or willet
Totanus semipalmatus
Willet
Catoptrophorus semipalmatus
14 ¾ × 20 ½

PLATE CCLXXV
Noddy tern
Sterna stolida
Brown noddy
Anous stolidus
12 ⅛ × 19 ¼

PLATE CCLXXVI
King duck
Fuligula spectabilis
King eider
Somateria spectabilis
25 ½ × 38 ⅛

PLATE CCLXXVII
Hutchins's barnacle goose
Anser hutchinsii
Canada goose
Branta canadensis
26 × 21 ⅞

PLATE CCLXXVIII
Schinz's sandpiper
Tringa schinzii
White-rumped sandpiper
Calidris fuscicollis
12 ⅛ × 19 ¼

PLATE CCLXXIX
Sandwich tern
Sterna boyssii
Sandwich tern
Sterna sandvicensis
12 ⅛ × 19 ¼

PLATE CCLXXX
Black tern
Sterna nigra
Black tern
Chlidonias niger
19 ⅝ × 12 ⅜

PLATE CCLXXXI
Great white heron
Ardea occidentalis
Great blue heron
Ardea herodias
25 ½ × 38 ⅛

PLATE CCLXXXII
White-winged silvery gull
Larus leucopterus
Iceland gull
Larus glaucoides
20 ⅝ × 25 ⅞

PLATE CCLXXXIII
Wandering shearwater
Puffinus cinereus
Great shearwater
Puffinus gravis
12 ⅜ × 19 ⅝

PLATE CCLXXXIV
Purple sandpiper
Tringa maritima
Purple sandpiper
Calidris maritima
12 ⅜ × 19 ⅝

PLATE CCLXXXV
Fork-tailed gull
Larus sabini
Sabine's gull and sanderling
Xema sabini and Calidris alba
12 ¼ × 19 ¼

PLATE CCLXXXVI
White-fronted goose
Anser albifrons
Greater white-fronted goose
Anser albifrons
25 ½ × 38 ⅛

PLATE CCLXXXVII
Ivory gull
Larus eburneus
Ivory gull
Pagophila eburnea
20 ⅝ × 30 ⅛

PLATE CCLXXXVIII
Yellow shank
Totanus flavipes
Lesser yellowlegs
Tringa flavipes
14 ⅝ × 20 ¼

PLATE CCLXXXIX
Solitary sandpiper
Totanus chloropygius
Solitary sandpiper
Tringa solitaria
12 ¼ × 19 ¼

PLATE CCXC
Red backed sandpiper
Tringa alpina
Dunlin
Calidris alpina
12 ⅛ × 19 ⅜

PLATE CCXCI
Herring gull
Larus argentatus
Herring gull
Larus argentatus
38 ⅛ × 25 ½

PLATE CCXCII
Crested grebe
Podiceps cristatus
Great crested grebe
Podiceps cristatus
20 ⅝ × 30 ⅛

PLATE CCXCIII
Large billed puffin
Mormon glacialis
Horned puffin
Fratercula corniculata
14 ¾ × 20 ½

PLATE CCXCIV
Pectoral sandpiper
Tringa pectoralis
Pectoral sandpiper
Calidris melanotos
12 ⅛ × 19 ¼

PLATE CCXCV
Manks shearwater
Puffinus anglorum
Manx shearwater
Puffinus puffinus
12 ¼ × 19 ¼

PLATE CCXCVI
Barnacle goose
Anser leucopsis
Barnacle goose
Branta leucopsis
25 ½ × 38 ⅛

PLATE CCXCVII
Harlequin duck
Fuligula histrionica
Harlequin duck
Histrionicus histrionicus
20 ⅝ × 30 ¼

PLATE CCXCVIII
Red-necked grebe
Podiceps rubricollis
Red-necked grebe
Podiceps grisegena
14 ¾ × 20 ⅜

PLATE CCXCIX
Dusky petrel
Puffinus obscurus
Audubon's shearwater
Puffinus lherminieri
12 ¼ × 19 ⅜

PLATE CCC
Golden plover
Charadrius pluvialis
American golden plover
Pluvialis dominica
14 ¾ × 20 ½

PLATE CCCI
Canvas backed duck
Fuligula vallisneria
Canvasback
Aythya valisineria
25 ½ × 38 ⅛

PLATE CCCII
Dusky duck
Anas obscura
American black duck
Anas rubripes
21 ⅛ × 30 ⅜

PLATE CCCIII
Bartram sandpiper
Totanus bartramius
Upland sandpiper
Bartramia longicauda
14 ⅞ × 21

PLATE CCCIV
Turn-stone
Strepsilas interpres
Ruddy turnstone
Arenaria interpres
14 ¾ × 21 ⅝

PLATE CCCV
Purple gallinule
Gallinula martinica
American purple gallinule
Porphyrio martinica
12 ⅜ × 19 ½

PLATE CCCVI
Great northern diver or loon
Colymbus glacialis
Great northern diver
Gavia immer
25 ⅝ × 38 ¼

PLATE CCCVII
Blue crane or heron
Ardea cœrulea
Little blue heron
Egretta caerulea
21 ¼ × 30 ½

PLATE CCCVIII
Tell-tale godwit or snipe
Totanus melanoleucus
Greater yellowlegs
Tringa melanoleuca
14 ⅞ × 21

PLATE CCCIX
Great tern
Sterna hirundo
Common tern
Sterna hirundo
19 ⅜ × 15 ¼

PLATE CCCX
Spotted sandpiper
Totanus macularius
Spotted sandpiper
Actitis macularius
14 ⅝ × 21 ⅛

PLATE CCCXI
American white pelican
Pelicanus americanus
American white pelican
Pelecanus erythrorhynchos
38 ¼ × 25 ¾

PLATE CCCXII
Long-tailed duck
Fuligula glacialis
Long-tailed duck
Clangula hyemalis
21 ¼ × 30 ¼

PLATE CCCXIII
Blue-winged teal
Anas discors
Blue-winged teal
Anas discors
14 ¾ × 20 ½

PLATE CCCXIV
Black-headed gull
Larus atricilla
Laughing gull
Larus atricilla
14 ¾ × 20 ½

PLATE CCCXV
Red-breasted sandpiper
Tringa islandica
Red knot
Calidris canutus
12 ¼ × 19 ½

PLATE CCCXVI
Black-bellied darter
Plotus anhinga
Anhinga
Anhinga anhinga
38 ¼ × 25 ⅝

PLATE CCCXVII
Black or surf duck
Fuligula perspicillata
Surf scoter
Melanitta perspicillata
21 ¼ × 30 ¼

PLATE CCCXVIII
American avocet
Recurvirostra americana
American avocet
Recurvirostra americana
14 ⅝ × 20 ⅜

PLATE CCCXIX
Lesser tern
Sterna minuta
Least tern
Sterna antillarum
19 ½ × 12 ¼

PLATE CCCXX
Little sandpiper
Tringa pusilla
Least sandpiper
Calidris minutilla
14 ¾ × 20 ½

PLATE CCCXXI
Roseate spoonbill
Platalea ajaja
Roseate spoonbill
Ajaia ajaja
25 ⅝ × 35 ¼

PLATE CCCXXII
Red-headed duck
Fuligula ferina
Redhead
Aythya americana
20 ⅝ × 26

PLATE CCCXXIII
Black skimmer or shearwater
Rhincops nigra
Black skimmer
Rynchops niger
21 × 21 ⅛

PLATE CCCXXIV
Bonapartian gull
Larus bonapartii
Bonaparte's gull
Larus philadelphia
21 ⅛ × 16 ⅞

PLATE CCCXXV
Buffel-headed duck
Fuligula albeola
Bufflehead
Bucephala albeola
14 ⅞ × 20 ½

PLATE CCCXXVI
Gannet
Sula bassana
Northern gannet
Morus bassanus
25 ⅝ × 38 ¼

PLATE CCCXXVII
Shoveller duck
Anas clypeata
Northern shoveler
Anas clypeata
21 × 30 ¼

PLATE CCCXXVIII
Long-legged avocet
Himantopus nigricollis
Black-winged stilt
Himantopus himantopus
14 ⅞ × 20 ½

PLATE CCCXXIX
Yellow-breasted rail
Rallus noveboracencis
Yellow rail
Coturnicops noveboracensis
12 ¼ × 19 ½

PLATE CCCXXX
Ring plover
Charadrius semipalmatus
Semipalmated plover
Charadrius semipalmatus
12 ¼ × 19 ½

PLATE CCCXXXI
Goosander
Mergus merganser
Goosander
Mergus merganser
25 ⅝ × 38 ¼

PLATE CCCXXXII
Pied duck
Fuligula labradora
Labrador duck
Camptorhynchus labradorius
21 ¼ × 30 ⅛

PLATE CCCXXXIII
Green heron
Ardea virescens
Green heron
Butorides virescens
20 ¼ × 22 ⅜

PLATE CCCXXXIV
Black-bellied plover
Charadrius helveticus
Grey plover
Pluvialis squatarola
15 ⅛ × 21

PLATE CCCXXXV
Red-breasted snipe
Scolopax grisea
Short-billed dowitcher
Limnodromus griseus
12 ¼ × 19 ⅜

PLATE CCCXXXVI
Yellow-crowned heron
Ardea violacea
Yellow-crowned night heron
Nyctanassa violacea
38 ¼ × 25 ⅝

PLATE CCCXXXVII
American bittern
Ardea minor
American bittern
Botaurus lentiginosus
22 ⅞ × 28 ⅛

PLATE CCCXXXVIII
Bemaculated duck
Anas glocitans
Mallard × gadwall
Anas platyrhynchos × *Anas strepera*
18 ½ × 23 ⅞

PLATE CCCXXXIX
Little auk
Uria alle
Little auk
Alle alle
12 ⅜ × 19 ⅞

PLATE CCCXL
Least stormy-petrel
Thalassidroma pelagica
European storm petrel
Hydrobates pelagicus
12 ½ × 19 ⅜

PLATE CCCXLI
Great auk
Alca impennis
Great auk
Pinguinus impennis
25 ⅝ × 38 ¼

PLATE CCCXLII
Golden-eye duck
Fuligula clangula
Common goldeneye
Bucephala clangula
21 ¼ × 30 ⅜

PLATE CCCXLIII
Ruddy duck
Fuligula rubida
Ruddy duck
Oxyura jamaicensis
16 × 26 ¼

PLATE CCCXLIV
Long-legged sandpiper
Tringa himantopus
Stilt-sandpiper
Calidris himantopus
12 ½ × 19 ⅞

PLATE CCCXLV
American widgeon
Anas americana
American wigeon
Anas americana
15 × 20

PLATE CCCXLVI
Black-throated diver
Colymbus arcticus
Black-throated diver
Gavia arctica
25 ⅝ × 38 ¼

PLATE CCCXLVII
Smew or white Nun
Mergus albellus
Smew
Mergus albellus
26 ⅜ × 22 ⅛

PLATE CCCXLVIII
Gadwall duck
Anas strepera
Gadwall
Anas strepera
16 ⅞ × 24 ⅞

PLATE CCCXLIX
Least water-hen
Rallus jamaicensis
Black rail
Laterallus jamaicensis
12 ¼ × 19 ½

PLATE CCCL
Rocky mountain plover
Charadrius montanus
Mountain plover
Charadrius montanus
12 ¼ × 19 ⅜

PLATE CCCLI
Great cinereous owl
Strix cinerea
Great grey owl
Strix nebulosa
38 ¼ × 25 ⅝

PLATE CCCLII
Black-winged hawk
Falco dispar
White-tailed kite
Elanus leucurus
30 ⅜ × 21 ¼

PLATE CCCLIII
Species 1 and 2
Chesnut-backed titmouse
Parus rufescens
Chesnut-backed chickadee
Parus rufescens

Species 3 and 4
Black-capt titmouse
Parus atricapillus
Black-capped chickadee
Parus atricapillus

Species 5 and 6
Chesnut-crowned titmouse
Parus minimus
Bushtit
Psaltriparus minimus
19 ⅞ × 14 ½

PLATE CCCLIV
Species 1 and 2
Louisiana tanager
Tanagra ludoviciana
Western tanager
Piranga ludoviciana

Species 3 and 4
Scarlet tanager
Tanagra rubra
Scarlet tanager
Piranga olivacea
12 ⅜ × 19 ⅞

PLATE CCCLV
MacGillivray's finch
Fringilla macgillivraii
Seaside sparrow
Ammodramus maritimus
19 ½ × 12 ⅜

PLATE CCCLVI
Marsh hawk
Falco cyaneus
Hen harrier
Circus cyaneus
38 ¼ × 25 ⅝

PLATE CCCLVII
American magpie
Corvus pica
Black-billed magpie
Pica hudsonica
25 ½ × 21 ¼

PLATE CCCLVIII
Pine grosbeak
Pyrrhula enucleator
Pine grosbeak
Pinicola enucleator
20 ½ × 14 ¾

PLATE CCCLIX
Species 1 and 2
Arkansaw flycatcher
Muscicapa verticalis
Say's phoebe
Sayornis saya

Species 3
Swallow tailed flycatcher
Muscicapa forficata
Scissor-tailed flycatcher
Tyrannus forficatus

Species 4 and 5
Says flycatcher
Muscicapa saya
Western kingbird
Tyrannus verticalis
21 ¾ × 14

PLATE CCCLX
Species 1, 2 and 3
Winter wren
Sylvia troglodytes
Winter wren
Troglodytes troglodytes

Species 4
Rock wren
Troglodytes obselata
Rock wren
Salpinctes obsoletus
19 ¾ × 12 ⅜

PLATE CCCLXI
Long-tailed or dusky grous
Tetrao obscurus
Blue grouse
Dendragapus obscurus
25 ⅝ × 38 ¼

PLATE CCCLXII
Species 1
Stellers jay
Corvus stellerii
Steller's jay
Cyanocitta stelleri

Species 2
Yellow billed magpie
Corvus nutallii
Yellow-billed magpie
Pica nuttalli

Species 3
Ultramarine jay
Corvus ultramarinus
Western scrub jay
Aphelocoma californica

Species 4 and 5
Clark's crow
Corvus columbianus
Clark's nutcracker
Nucifraga columbiana
26 ⅛ × 21 ⅞

PLATE CCCLXIII
Bohemian chatterer
Bombycilla garrula
Bohemian waxwing
Bombycilla garrulus
19 ¾ × 12 ½

PLATE CCCLXIV
White-winged crossbill
Loxia leucoptera
White-winged crossbill
Loxia leucoptera
19 ⅝ × 12 ⅜

PLATE CCCLXV
Lapland long-spur
Fringilla laponica
Lapland longspur
Calcarius lapponicus
12 ⅜ × 19 ⅝

PLATE CCCLXVI
Iceland or jer falcon
Falco islandicus
Gyrfalcon
Falco rusticolus
38 ¼ × 25 ⅝

PLATE CCCLXVII
Band-tailed pigeon
Columba fasciata
Band-tailed pigeon
Columba fasciata
29 ¼ × 21 ¾

PLATE CCCLXVIII
Rock grous
Tetrao rupestris
Rock ptarmigan
Lagopus muta
16 ¼ × 21 ⅜

PLATE CCCLXIX
Species 1
Mountain mocking bird
Orpheus montanus
Sage thrasher
Oreoscoptes montanus

Species 2
Varied thrush
Turdus nævius
Varied thrush
Ixoreus naevius
19 ¾ × 14 ½

PLATE CCCLXX
American water ouzel
Cinclus americanus
American dipper
Cinclus mexicanus
12 ⅜ × 19 ¾

PLATE CCCLXXI
Cock of the plains
Tetrao urophasianus
Greater sage grouse
Centrocercus urophasianus
25 ⅝ × 38 ¼

PLATE CCCLXXII
Common buzzard
Buteo vulgaris
Swainson's hawk
Buteo swainsoni
27 ⅜ × 23 ¾

PLATE CCCLXXIII
Species 1
Evening grosbeak
Fringilla vespertina
Evening grosbeak
Coccothraustes vespertinus

Species 2
Spotted grosbeak
Fringilla maculata
Black-headed grosbeak
Pheucticus melanocephalus
19 ½ × 12 ⅜

PLATE CCCLXXIV
Sharp-shinned hawk
Falco velox
Sharp-shinned hawk
Accipiter striatus
19 ½ × 14 ¾

PLATE CCCLXXV
Lesser red-poll
Fringilla linaria
Common redpoll
Carduelis flammea
19 ½ × 12 ⅜

PLATE CCCLXXVI
Trumpeter swan
Cygnus buccinator
Trumpeter swan
Cygnus buccinator
25 ⅝ × 38 ¼

PLATE CCCLXXVII
Scolopaceus courlan
Aramus scolopaceus
Limpkin
Aramus guarauna
21 × 33 ¼

PLATE CCCLXXVIII
Hawk owl
Strix funerea
Northern hawk-owl
Surnia ulula
26 ⅛ × 21 ⅝

PLATE CCCLXXIX
Ruff-necked humming-bird
Trochilus rufus
Rufous hummingbird
Selasphorus rufus
19 ⅝ × 12 ⅜

PLATE CCCLXXX
Tengmalm's owl
Strix tengmalmi
Tengmalm's owl
Aegolius funereus
20 ¾ × 15 ¾

PLATE CCCLXXXI
Snow goose
Anser hyperboreus
Snow goose
Anser caerulescens
25 ⅝ × 38 ¼

PLATE CCCLXXXII
Sharp-tailed grous
Tetrao phasianellus
Sharp-tailed grouse
Tympanuchus phasianellus
21 ⅝ × 29 ¼

PLATE CCCLXXXIII
Long-eared owl
Strix otus
Long-eared owl
Asio otus
19 ¾ × 12 ½

PLATE CCCLXXXIV
Black-throated bunting
Fringilla americana
Dickcissel
Spiza americana
19 ⅝ × 12 ⅜

PLATE CCCLXXXV
Species 1
Bank swallow
Hirundo riparia
Collared sand martin
Riparia riparia

Species 2
Violet-green swallow
Hirundo thalassinus
Violet-green swallow
Tachycineta thalassina
19 ⅛ × 15 ⅞

PLATE CCCLXXXVI
White heron
Ardea alba
Great egret
Ardea alba
25 ⅝ × 38 ¼

PLATE CCCLXXXVII
Glossy ibis
Ibis falcinellus
Glossy ibis
Plegadis falcinellus
21 ⅞ × 26 ⅛

PLATE CCCLXXXVIII
Species 1
Nuttall's starling
Icterus tricolor
Tricoloured blackbird
Agelaius tricolor

Species 2
Yellow-headed troopial
Icterus xanthocephalus
Yellow-headed blackbird
Xanthocephalus xanthocephalus

Species 3
Bullock's oriole
Icterus bullockii
Bullock's oriole
Icterus bullockii
19 ¾ × 12 ⅜

PLATE CCCLXXXIX
Red-cockaded woodpecker
Picus querulus
Red-cockaded woodpecker
Picoides borealis
19 ¾ × 12 ¼

PLATE CCCXC
Species 1
Lark finch
Fringilla grammaca
Lark-sparrow
Chondestes grammacus

Species 2
Prairie finch
Fringilla bicolor
Lark-bunting
Calamospiza melanocorys

Species 3
Brown song sparrow
Fringilla cinerea
Song sparrow
Melospiza melodia
19 ⅝ × 12 ¼

PLATE CCCXCI
Brant goose
Anser bernicla
Brent goose
Branta bernicla
25 ⅝ × 38 ¼

PLATE CCCXCII
Louisiana hawk
Buteo harrisi
Harris's hawk
Parabuteo unicinctus
21 ⅞ × 26

PLATE CCCXCIII
Species 1
Townsend's warbler
Sylvia townsendi
Townsend's warbler
Dendroica townsendi

Species 2
Arctic blue-bird
Sialia arctica
Mountain bluebird
Sialia currucoides

Species 3
Western blue-bird
Sialia occidentalis
Western bluebird
Sialia mexicana
19 ½ × 12 ⅜

PLATE CCCXCIV
Species 1
Chestnut-coloured finch
Plectrophanes ornata
Chestnut-collared longspur
Calcarius ornatus

Species 2
Black-headed siskin
Fringilla magellanica
Hooded siskin
Carduelis magellanica

Species 3
Black crown bunting
Emberiza atricapilla
Golden-crowned sparrow
Zonotrichia atricapilla

Species 4
Arctic ground finch
Pipilo arctica
Spotted towhee
Pipilo maculatus
19 ⅝ × 12 ¼

PLATE CCCXCV
Species 1
Audubon's warbler
Sylvia auduboni
Yellow-rumped warbler
Dendroica coronata

Species 2
Hermit warbler
Sylvia occidentalis
Hermit warbler
Dendroica occidentalis

Species 3
Black-throated gray warbler
Sylvia nigrescens
Black-throated grey warbler
Dendroica nigrescens
19 ⅝ × 12 ¼

PLATE CCCXCVI
Burgomaster gull
Larus glaucus
Glaucous gull
Larus hyperboreus
25 ⅝ × 38 ¼

PLATE CCCXCVII
Scarlet ibis
Ibis rubra
Scarlet ibis
Eudocimus ruber
21 ⅝ × 29 ⅜

PLATE CCCXCVIII
Species 1
Lazuli finch
Fringilla amœna
Lazuli bunting
Passerina amoena

Species 2
Clay-coloured finch
Fringilla pallida
Clay-coloured sparrow
Spizella pallida

Species 3
Oregon snow finch
Fringilla oregona
Dark-eyed junco
Junco hyemalis
19 ½ × 12 ¼

PLATE CCCXCIX
Species 1
Black-throated green warbler
Sylvia virens
Black-throated green warbler
Dendroica virens

Species 2
Blackburnian warbler
Sylvia blackburniæ
Blackburnian warbler
Dendroica fusca

Species 3
Mourning warbler
Sylvia philadelphia
MacGillivray's warbler
Oporornis tolmiei
19 ½ × 12 ¼

PLATE CCCC
Species 1
Arkansaw siskin
Fringilla spaltria
Lesser goldfinch
Carduelis psaltria

Species 2
Mealy red-poll
Linota borealis
Hoary redpoll
Carduelis hornemanni

Species 3
Louisiana tanager
Tanagra ludoviciana
Western tanager
Piranga ludoviciana

Species 4
Townsend's finch
Emberiza townsendi
Townsend's bunting
Emberiza townsendi

Species 5
Buff-breasted finch
Emberiza picta
Smith's longspur
Calcarius pictus
20 × 12 ½

PLATE CCCCI
Red-breasted merganser
Mergus serrator
Red-breasted merganser
Mergus serrator
25 ⅝ × 38 ¼

PLATE CCCCII
Species 1 and 2
Black-throated guillemot
Mergulus antiquus
Ancient murrelet and Kittlitz's murrelet
Synthliboramphus antiquus and Brachyramphus brevirostris

Species 3
Nobbed-billed auk
Phaleris nodirostris
Least auklet
Aethia pusilla

Species 4
Curled-crested auk
Phaleris superciliata
Crested auklet
Aethia cristatella

Species 5
Horned-billed guillemot
Ceratorrhina occidentalis
Rhinoceros auklet
Cerorhinca monocerata
18 ⅝ × 28

PLATE CCCCIII
Golden-eye duck
Clangula vulgaris
Barrow's goldeneye
Bucephala islandica
12 ¼ × 19 ½

PLATE CCCCIV
Eared grebe
Podiceps auritus
Black-necked grebe
Podiceps nigricollis
12 ⅜ × 19 ⅝

PLATE CCCCV
Semipalmated sandpiper
Tringa semipalmata
Semipalmated sandpiper
Calidris pusilla
12 ⅜ × 19 ½

PLATE CCCCVI
Trumpeter swan
Cygnus buccinator
Trumpeter swan
Cygnus buccinator
25 ⅝ × 38 ¼

PLATE CCCCVII
Dusky albatros
Diomedea fusca
Sooty albatross
Phoebetria fusca
21 ⅝ × 28 ⅜

PLATE CCCCVIII
American scoter duck
Fuligula americana
Black scoter
Melanitta nigra
16 ⅝ × 21 ⅞

PLATE CCCCIX
Species 1
Havell's tern
Sterna havelli
Forster's tern
Sterna forsteri

Species 2
Trudeau's tern
Sterna trudeaui
Trudeau's tern
Sterna trudeaui
15 ¼ × 24 ¾

PLATE CCCCX
Marsh tern
Sterna anglica
Gull-billed tern
Sterna nilotica
19 ⅜ × 15 ⅞

PLATE CCCCXI
Common american swan
Cygnus americanus
Tundra swan
Cygnus columbianus
25 ⅝ × 38 ¼

PLATE CCCCXII
Species 1
Violet-green cormorant
Phalacrocorax resplendens
Pelagic cormorant
Phalacrocorax pelagicus
23 ¾ × 27 ½

Species 2
Townsend's cormorant
Phalacrocorax townsendi
Brandt's cormorant
Phalacrocorax penicillatus

PLATE CCCCXIII
Californian partridge
Perdix californica
Californian quail
Callipepla californica
12 ¼ × 19 ½

PLATE CCCCXIV
Species 1
Golden-winged warbler
Sylvia chrysoptera
Golden-winged warbler
Vermivora chrysoptera

Species 2
Cape may warbler
Sylvia maritima
Cape may warbler
Dendroica tigrina
19 ½ × 12 ¼

PLATE CCCCXV
Species 1
Brown creeper
Certhia familiaris
American treecreeper
Certhia americana

Species 2
Californian nuthatch
Sitta pygmea
Pygmy nuthatch
Sitta pygmaea
19 ½ × 12 ½

PLATE CCCCXVI
Species 1 and 2
Hairy woodpecker
Picus villosus
Hairy woodpecker
Picoides villosus

Species 3 and 4
Red-bellied woodpecker
Picus carolinus
Red-bellied woodpecker
Melanerpes carolinus

Species 5 and 6
Red-shafted woodpecker
Picus mexicanus
Northern flicker
Colaptes auratus

Species 7 and 8
Lewis' woodpecker
Picus torquatus
Lewis's woodpecker
Melanerpes lewis

Species 9 and 10
Red-breasted woodpecker
Picus ruber
Red-breasted sapsucker
Sphyrapicus ruber
38 ¼ × 25 ⅝

PLATE CCCCXVII
Species 1 and 2
Maria's woodpecker
Picus martini
Hairy woodpecker
Picoides villosus

Species 3 and 4
Three-toed woodpecker
Picus hirsitus
Three-toed woodpecker
Picoides tridactylus

Species 5 and 6
Phillips' woodpecker
Picus phillipsi
Hairy woodpecker
Picoides villosus

Species 7
Canadian woodpecker
Picus canadensis
Hairy woodpecker
Picoides villosus

Species 8 and 9
Harris's woodpecker
Picus harrisi
Hairy woodpecker
Picoides villosus

Species 10
Audubon's woodpecker
Picus auduboni
Hairy woodpecker
Picoides villosus
30 ½ × 22 ½

PLATE CCCCXVIII
Species 1
American ptarmigan
Tetrao mutus
Rock ptarmigan
Lagopus muta

Species 2
White-tailed grous
Tetrao leucurus
White-tailed ptarmigan
Lagopus leucura
16 ¾ × 22 ⅞

PLATE CCCCXIX
Species 1
Little tawny thrush
Turdus minor
Hermit thrush
Catharus guttatus

Species 2
[No common name given]
Ptiliogony's townsendi
Townsend's solitaire
Myadestes townsendi

Species 3
Canada jay
Corvus canadensis
Grey jay
Perisoreus canadensis
19 ½ × 12 ⅜

PLATE CCCCXX
Prairie starling
Icterus gubernator
Red-winged blackbird
Agelaius phoeniceus
19 ⅜ × 12 ¼

PLATE CCCCXXI
Brown pelican
Pelicanus fuscus
Brown pelican
Pelecanus occidentalis
25 ⅝ × 38 ¼

PLATE CCCCXXII
Rough-legged falcon
Buteo lagopus
Rough-legged buzzard
Buteo lagopus
38 ¼ × 25 ⅝

PLATE CCCCXXIII
Species 1
Thick-legged partridge
Perdix neoxenus
Crested bobwhite
Colinus cristatus

Species 2
Plumed partridge
Perdix plumifera
Mountain quail
Oreortyx pictus
12 ¾ × 21 ⅜

PLATE CCCCXXIV
Species 1
Lazuli finch
Fringilla amœna
Lazuli bunting
Passerina amoena

Species 2
Crimson-necked bull-finch
Pyrrhula frontalis
House finch
Carpodacus mexicanus

Species 3
Grey-crowned linnet
Linaria tephrocotis
Grey-crowned rosy finch
Leucosticte tephrocotis

Species 4
Cow-pen bird
Icterus pecoris
Brown-headed cowbird
Molothrus ater

Species 5
Evening grosbeak
Fringilla vespertina
Evening grosbeak
Coccothraustes vespertinus

Species 6
Brown longspur
Plectrophanes townsendi
Fox sparrow
Passerella iliaca
20 × 13 ¼

PLATE CCCCXXV
Columbian humming bird
Trochilus anna
Anna's hummingbird
Calypte anna
19 ½ × 12 ¼

PLATE CCCCXXVI
Californian vulture
Cathartes californianus
California condor
Gymnogyps californianus
38 ¼ × 25 ⅝

PLATE CCCCXXVII
Species 1
White-legged oyster-catcher
Hæmatopus bachmani
American black oyster-catcher
Haematopus bachmani

Species 2
Slender-billed oyster-catcher
Hœmatopus townsendi
American black oyster-catcher
Haematopus bachmani
21 ¾ × 27 ¼

PLATE CCCCXXVIII
Townsend's sandpiper
Frinca townsendi
Surfbird
Aphriza virgata
18 ⅝ × 16 ⅛

PLATE CCCCXXIX
Western duck
Fuligula stelleri
Steller's eider
Polysticta stelleri
12 ¾ × 21 ⅜

PLATE CCCCXXX
Slender-billed guillemot
Uria townsendi
Marbled murrelet
Brachyramphus marmoratus
12 ¼ × 19 ⅛

PLATE CCCCXXXI
American flamingo
Phœnicopterus ruber
Greater flamingo
Phoenicopterus ruber
38 ¼ × 25 ⅝

PLATE CCCCXXXII
Species 1
Burrowing owl
Strix cunicularia
Burrowing owl
Athene cunicularia

Species 2
Large-headed burrowing owl
Strix californica
Burrowing owl
Athene cunicularia

Species 3
Little night owl
Strix noctua
Little owl
Athene noctua

Species 4 and 5
Columbian owl
Strix passerinoides
Northern pygmy owl
Glaucidium gnoma

Species 6
Short-eared owl
Strix brachyotus
Short-eared owl
Asio flammeus
21 ¾ × 26

PLATE CCCCXXXIII
Species 1
Bullock's oriole
Icterus bullocki
Bullock's oriole
Icterus bullockii

Species 2
Baltimore oriole
Icterus baltimore
Baltimore oriole
Icterus galbula

Species 3
Mexican goldfinch
Carduellis mexicanus
Lesser goldfinch
Carduelis psaltria

Species 4
Varied thrush
Turdus nœvius
Varied thrush
Ixoreus naevius

Species 5
Common water thrush
Turdus aquaticus
Northern waterthrush
Seiurus noveboracensis
20 ⅜ × 14 ⅛

PLATE CCCCXXXIV
Species 1
Little tyrant fly-catcher
Tyrannula pusilla
Least flycatcher
Empidonax minimus

Species 2
Small-headed fly-catcher
Muscicapa minuta
Small-headed flycatcher
Sylvania microcephala

Species 3
Blue mountain warbler
Sylvia montana
Blue mountain warbler
Sylvia montana

Species 4
Bartram's vireo
Vireo bartrami
Red-eyed vireo
Vireo olivaceus

Species 5
Short-legged pewee
Muscicapa phœbe
Western wood pewee
Contopus sordidulus

Species 6
Rocky mountain fly-catcher
Tyrannula nigricans
Black phoebe
Sayornis nigricans
19 ½ × 12 ¼

PLATE CCCCXXXV
Species 1
Columbian water ouzel
Cinclus townsendi
American dipper
Cinclus mexicanus

Species 2
Arctic water ouzel
Cinclus mortoni
American dipper
Cinclus mexicanus
19 ½ × 12 ¼

索 引

（英国自然博物馆给出的英文名，中文名，图版编号）

Acadian flycatcher, 绿纹霸鹟, PLATE CXLIV
American avocet, 褐胸反嘴鹬, PLATE CCCXVIII
American bittern, 美洲麻鳽, PLATE CCCXXXVII
American black duck, 北美黑鸭, PLATE CCCII
American black oyster-catcher, 北美蛎鹬, PLATE CCCCXXVII
American coot, 美洲骨顶, PLATE CCXXXIX
American crow, 短嘴鸦, PLATE CLVI
American dipper, 美洲河乌, PLATES CCCLXX, CCCCXXXV
American golden plover, 美洲金鸻, PLATE CCC
American goldfinch, 美洲金翅雀, PLATE 33
American kestrel, 美洲隼, PLATE CXLII
American oystercatcher, 美洲蛎鹬, PLATE CCXXIII
American purple gallinule, 紫青水鸡, PLATE CCCV
American redstart, 橙尾鸲莺, PLATE 40
American robin, 旅鸫, PLATE CXXXI
American treecreeper, 褐头鸸, PLATE CCCCXV
American tree sparrow, 美洲树雀鹀, PLATE CLXXXVIII
American white pelican, 美洲鹈鹕, PLATE CCCXI
American wigeon, 绿眉鸭, PLATE CCCXLV
American woodcock, 小丘鹬, PLATE CCLXVIII
Anna's hummingbird, 安氏蜂鸟, PLATE CCCCXXV
Ancient murrelet, 扁嘴海雀, PLATE CCCCII
Anhinga, 美洲蛇鹈, PLATE CCCXVI
Arctic skua, 短尾贼鸥, PLATE CCLXXII
Arctic tern, 北极燕鸥, PLATE CCL
Atlantic puffin, 北极海鹦, PLATE CCXIII
Audubon's shearwater, 奥氏鹱, PLATE CCXCIX

Bachman's sparrow, 巴氏猛雀鹀, PLATE CLXV
Bachman's warbler, 黑胸虫森莺, PLATE CLXXXV
Bald eagle, 白头海雕, PLATES 11, 31, CXXVI
Baltimore oriole, 橙腹拟鹂, PLATES 12, CCCCXXXIII
Band-tailed pigeon, 斑尾鸽, PLATE CCCLXVII
Barn owl, 仓鸮, PLATE CLXXI
Barn swallow, 家燕, PLATE CLXXIII
Barnacle goose, 白颊黑雁, PLATE CCXCVI
Barred owl, 横斑林鸮, PLATE 46
Barrow's goldeneye, 巴氏鹊鸭, PLATE CCCCIII
Bay-breasted warbler, 栗胸林莺, PLATES 69, 88
Belted kingfisher, 白腹鱼狗, PLATE 77
Bewick's wren, 比氏苇鹪鹩, PLATE 18
Black guillemot, 白翅斑海鸽, PLATE CCXIX
Black phoebe, 黑长尾霸鹟, PLATE CCCCXXXIV
Black rail, 黑田鸡, PLATE CCCXLIX
Black scoter, 黑海番鸭, PLATE CCCCVIII
Black skimmer, 黑剪嘴鸥, PLATE CCCXXIII
Black tern, 黑浮鸥, PLATE CCLXXX
Black vulture, 黑头美洲鹫, PLATE CVI
Black-and-white warbler, 黑白森莺, PLATE 90
Black-backed woodpecker, 黑背啄木鸟, PLATE CXXXII
Black-billed cuckoo, 黑嘴美洲鹃, PLATE 32
Black-billed magpie, 黑嘴喜鹊, PLATE CCCLVII
Blackburnian warbler, 橙胸林莺, PLATES CCCXCIX, CXXXIV, CXXXV
Black-capped chickadee, 黑顶山雀, PLATE CCCLIII
Black-crowned night heron, 夜鹭, PLATE CCXXXVI
Black-headed grosbeak, 黑头白斑翅雀, PLATE CCCLXXIII
Black-legged kittiwake, 三趾鸥, PLATE CCXXIV
Black-necked grebe, 黑颈䴙䴘, PLATE CCCCIV
Blackpoll warbler, 白颊林莺, PLATE CXXXIII
Black-throated blue warbler, 黑喉蓝林莺, PLATES CLV, CXLVIII
Black-throated diver, 黑喉潜鸟, PLATE CCCXLVI
Black-throated green warbler, 黑喉绿林莺, PLATE CCCXCIX
Black-throated grey warbler, 黑喉灰林莺, PLATE CCCXCV
Black-throated magpie-jay, 白喉鹊鸦, PLATE 96
Black-throated mango, 黑喉芒果蜂鸟, PLATE CLXXXIV
Black-winged stilt, 黑颈长脚鹬, PLATE CCCXXVIII
Blue grosbeak, 斑翅蓝彩鹀, PLATE CXXII
Blue grouse, 蓝镰翅鸡, PLATE CCCLXI
Blue jay, 冠蓝鸦, PLATE CII
Blue mountain warbler, 蓝山莺, PLATE CCCCXXXIV
Blue-grey gnatcatcher, 灰蓝蚋莺, PLATE 84
Blue-headed quail-dove, 蓝头鹑鸠, PLATE CLXXII
Blue-headed vireo, 蓝头莺雀, PLATE 28
Blue-winged teal, 蓝翅鸭, PLATE CCCXIII
Blue-winged warbler, 蓝翅虫森莺, PLATE 20
Boat-tailed grackle, 宽尾拟八哥, PLATE CLXXXVII
Bobolink, 刺歌雀, PLATE 54
Bohemian waxwing, 太平鸟, PLATE CCCLXIII
Bonaparte's gull, 博氏鸥, PLATE CCCXXIV
Boreal chickadee, 北山雀, PLATE CXCIV
Brandt's cormorant, 加州鸬鹚, PLATE CCCCXII
Brent goose, 黑雁, PLATE CCCXCI
Broad-winged hawk, 巨翅鵟, PLATE 91
Brown booby, 褐鲣鸟, PLATE CCVII
Brown noddy, 白顶玄燕鸥, PLATE CCLXXV
Brown pelican, 褐鹈鹕, PLATES CCLI, CCCCXXI

Brown thrasher, 褐弯嘴嘲鸫, PLATE CXVI
Brown-headed cowbird, 褐头牛鹂, PLATES 99, CCCCXXIV
Brown-headed nuthatch, 褐头䴓, PLATE CXXV
Brunnich's guillemot, 厚嘴崖海鸦, PLATE CCXLV
Buff-breasted sandpiper, 黄胸鹬, PLATE CCLXV
Bufflehead, 白枕鹊鸭, PLATE CCCXXV
Bullock's oriole, 布氏拟鹂, PLATES CCCLXXXVIII, CCCCXXXIII
Burrowing owl, 穴小鸮, PLATE CCCCXXXII
Bushtit, 短嘴长尾山雀, PLATE CCCLIII

Californian condor, 加州神鹫, PLATE CCCCXXVI
Californian quail, 珠颈斑鹑, PLATE CCCCXIII
Canada goose, 加拿大黑雁, PLATES CCI, CCLXXVII
Canada warbler, 加拿大威森莺, PLATES V, CIII
Canvasback, 帆背潜鸭, PLATE CCCI
Cape may warbler, 栗颊林莺, PLATE CCCCXIV
Carbonated warbler, PLATE 60
Carolina chickadee, 卡罗山雀, PLATE CLX
Carolina parakeet, 卡罗莱纳（长尾）鹦鹉, PLATE 26
Carolina wren, 卡罗苇鹪鹩, PLATE 78
Cedar waxwing, 雪松太平鸟, PLATE 43
Cerulean warbler, 白喉林莺, PLATES 48, 49
Chesnut-backed chickadee, 栗背山雀, PLATE CCCLIII
Chestnut-collared longspur, 栗领铁爪鹀, PLATE CCCXCIV
Chestnut-sided warbler, 栗胁林莺, PLATE 59
Chimney swift, 烟囱雨燕, PLATE CLVIII
Chipping sparrow, 棕顶雀鹀, PLATE CIV
Chuck-will's widow, 卡氏夜鹰, PLATE 52
Clapper rail, 长嘴秧鸡, PLATE CCIV
Clark's nutcracker, 北美星鸦, PLATE CCCLXII
Clay-coloured sparrow, 褐雀鹀, PLATE CCCXCVIII
Cliff swallow, 美洲燕, PLATE 68
Collared sand martin, 崖沙燕, PLATE CCCLXXXV
Common eider, 欧绒鸭, PLATE CCXLVI
Common goldeneye, 鹊鸭, PLATE CCCXLII
Common grackle, 拟八哥, PLATE VII
Common greenshank, 青脚鹬, PLATE CCLXIX
Common ground dove,地鸠, PLATE CLXXXII
Common moorhen, 黑水鸡, PLATE CCXLIV
Common nighthawk, 美洲夜鹰, PLATE CXLVII
Common raven, 渡鸦, PLATE CI
Common redpoll, 白腰朱顶雀, PLATE CCCLXXV
Common snipe, 扇尾沙锥, PLATE CCXLIII
Common tern, 普通燕鸥, PLATE CCCIX
Common yellowthroat, 黄喉地莺, PLATES 23, 24
Connecticut warbler, 灰喉地莺, PLATE CXXXVIII
Cooper's hawk, 库氏鹰, PLATES 36, CXLI
Crested auklet, 凤头海雀, PLATE CCCCII
Crested bobwhite, 冠齿鹑, PLATE CCCCXXIII
Crested caracara, 凤头巨隼, PLATE CLXI
Curlew sandpiper, 弯嘴滨鹬, PLATE CCLXIII
Cuvier's kinglet, 红冠戴菊, PLATE 55

Dark-eyed junco, 暗眼灯草鹀, PLATES 13, CCCXCVIII
Dickcissel, 美洲雀, PLATE CCCLXXXIV
Double-crested cormorant, 角鸬鹚, PLATES CCLII, CCLVII
Downy woodpecker, 绒啄木鸟, PLATE CXII
Dunlin, 黑腹滨鹬, PLATE CCXC

Eastern bluebird, 东蓝鸲, PLATE CXIII
Eastern kingbird, 东王霸鹟, PLATE 79
Eastern meadowlark, 东草地鹨, PLATE CXXXVI
Eastern phoebe, 灰胸长尾霸鹟, PLATE CXX
Eastern screech owl, 东美角鸮, PLATE 97
Eastern towhee, 棕胁唧鹀, PLATE 29
Eastern wood pewee, 东绿霸鹟, PLATE CXV
Eskimo curlew, 极北杓鹬, PLATE CCVIII
European storm petrel, 暴风海燕, PLATE CCCXL
Evening grosbeak, 黄昏锡嘴雀, PLATES CCCLXXIII, CCCCXXIV

Field sparrow, 田雀鹀, PLATE CXXXIX
Fish crow, 鱼鸦, PLATE CXLVI
Florida scrub jay, 丛鸦, PLATE 87
Fork-tailed flycatcher, 叉尾王霸鹟, PLATE CLXVIII
Forster's tern, 弗氏燕鸥, PLATE CCCCIX
Fox sparrow, 狐色雀鹀, PLATES CVIII, CCCCXXIV

Gadwall, 赤膀鸭, PLATE CCCXLVIII
Glaucous gull, 北极鸥, PLATE CCCXCVI
Glossy ibis, 彩鹮, PLATE CCCLXXXVII
Golden eagle, 金雕, PLATE CLXXXI
Golden-crowned kinglet, 金冠戴菊, PLATE CLXXXIII
Golden-crowned sparrow, 金冠带鹀, PLATE CCCXCIV
Golden-winged warbler, 金翅虫森莺, PLATE CCCCXIV
Goosander, 普通秋沙鸭, PLATE CCCXXXI
Grasshopper sparrow, 黄胸草鹀, PLATE CXXX
Gray catbird, 灰嘲鸫, PLATE CXXVIII
Great auk, 大海雀（已灭绝）, PLATE CCCXLI
Great black-backed gull, 大黑背鸥, PLATE CCXLI
Great blue heron, 大蓝鹭, PLATES CCXI, CCLXXXI
Great cormorant, 普通鸬鹚, PLATE CCLXVI
Great crested flycatcher, 大冠蝇霸鹟, PLATE CXXIX
Great crested grebe, 凤头䴙䴘, PLATE CCXCII
Great egret, 大白鹭, PLATE CCCLXXXVI
Great grey owl, 乌林鸮, PLATE CCCLI
Great grey shrike, 灰伯劳, PLATE CXCII
Great horned owl, 美洲雕鸮, PLATE 61
Great northern diver, 普通潜鸟, PLATE CCCVI
Great shearwater, 大鹱, PLATE CCLXXXIII
Greater flamingo, 大红鹳, PLATE CCCCXXXI
Greater prairie chicken, 草原松鸡, PLATE CLXXXVI
Greater sage grouse, 艾草松鸡, PLATE CCCLXXI
Greater scaup, 小潜鸭, PLATE CCXXIX
Greater white-fronted goose, 白额雁, PLATE CCLXXXVI
Greater yellowlegs, 大黄脚鹬, PLATE CCCVIII
Green heron, 美洲绿鹭, PLATE CCCXXXIII
Green-winged teal, 绿翅鸭, PLATE CCXXVIII
Grey jay, 灰噪鸦, PLATES CVII, CCCCXIX
Grey kingbird, 灰王霸鹟, PLATE CLXX
Grey phalarope, 灰瓣蹼鹬, PLATE CCLV
Grey plover, 灰鸻, PLATE CCCXXXIV
Grey-crowned rosy finch, 褐头牛鹂, PLATE CCCCXXIV
Guillemot, 崖海鸦, PLATE CCXVIII
Gull-billed tern, 鸥嘴噪鸥, PLATE CCCCX
Gyrfalcon, 矛隼, PLATES CXCVI, CCCLXVI

Hairy woodpecker, 长嘴啄木鸟, PLATES CCCCXVI, CCCCXVII
Harlequin duck, 丑鸭, PLATE CCXCVII
Harris's hawk, 栗翅鹰, PLATE CCCXCII
Hen harrier, 白尾鹞, PLATE CCCLVI

Henslow's sparrow, 亨氏草鹀, PLATE 70
Hermit thrush, 隐夜鸫, PLATES 58, CCCCXIX
Hermit warbler, 黄脸林莺, PLATE CCCXCV
Herring gull, 银鸥, PLATE CCXCI
Hoary redpoll, 白腰朱顶雀, PLATE CCCC
Hooded merganser, 棕胁秋沙鸭, PLATE CCXXXII
Hooded siskin, 黑头白斑翅雀, PLATE CCCXCIV
Hooded warbler, 黑枕威森莺, PLATES IX, CX
Horned grebe, 角䴙䴘, PLATE CCLIX
Horned lark, 角百灵, PLATE CC
Horned puffin, 角海鹦, PLATE CCXCIII
House finch, 家朱雀, PLATE CCCCXXIV
House wren, 莺鹪鹩, PLATES 83, CLXXIX
Hudsonian godwit, 棕塍鹬, PLATE CCLVIII

Iceland gull, 冰岛鸥, PLATE CCLXXXII
Indigo bunting, 靛彩鹀, PLATE 74
Ivory gull, 白鸥, PLATE CCLXXXVII
Ivory-billed woodpecker, 象牙喙啄木鸟, PLATE 66

Kentucky warbler, 黄腹地莺, PLATE 38
Key West quail-dove, 绿顶鹑鸠, PLATE CLXVII
Killdeer, 双领鸻, PLATE CCXXV
King eider, 王绒鸭, PLATE CCLXXVI
King rail, 王秧鸡, PLATE CCIII
Kittlitz's murrelet, 扁嘴海雀, PLATE CCCCII
Labrador duck, 拉布拉多鸭（已灭绝）, PLATE CCCXXXII
Lapland longspur, 铁爪鹀, PLATE CCCLXV
Lark-bunting, 白斑黑鹀, PLATE CCCXC
Lark-sparrow, 鹨雀鹀, PLATE CCCXC
Laughing gull, 笑鸥, PLATE CCCXIV
Lazuli bunting, 白腹蓝彩鹀, PLATES CCCXCVIII, CCCCXXIV
Leach's storm petrel, 白腰叉尾海燕, PLATE CCLX
Least auklet, 小海雀, PLATE CCCCII
Least bittern, 姬苇鳽, PLATE CCX
Least flycatcher, 小纹霸鹟, PLATE CCCCXXXIV
Least sandpiper, 美洲小滨鹬, PLATE CCCXX
Least tern, 小白额燕鸥, PLATE CCCXIX
Lesser goldfinch, 暗背金翅雀, PLATES CCCC, CCCCXXXIII
Lesser yellowlegs, 小黄脚鹬, PLATE CCLXXXVIII
Lewis's woodpecker, 刘氏啄木鸟, PLATE CCCCXVI
Limpkin, 秧鹤, PLATE CCCLXXVII
Lincoln's sparrow, 林氏带鹀, PLATE CXCIII
Little auk, 侏海雀, PLATE CCCXXXIX
Little blue heron, 小蓝鹭, PLATE CCCVII
Little owl, 纵纹腹小鸮, PLATE CCCCXXXII
Loggerhead shrike, 呆头伯劳, PLATE 57
Long-billed curlew, 长嘴杓鹬, PLATE CCXXXI
Long-eared owl, 长耳鸮, PLATE CCCLXXXIII
Long-tailed duck, 长尾鸭, PLATE CCCXII
Long-tailed skua, 长尾贼鸥, PLATE CCLXVII
Louisiana waterthrush, 白眉灶莺, PLATE 19

MacGillivray's warbler, 灰头地莺, PLATE CCCXCIX
Magnificent frigatebird, 华丽军舰鸟, PLATE CCLXXI
Magnolia warbler, 纹胸林莺, PLATES 50, CXXIII
Mallard x gadwall, 布氏鸭, PLATE CCCXXXVIII
Mallard, 绿头鸭, PLATE CCXXI
Mangrove cuckoo, 红树美洲鹃, PLATE CLXIX
Manx shearwater, 大西洋鹱, PLATE CCXCV
Marbled godwit, 云斑塍鹬, PLATE CCXXXVIII
Marbled murrelet, 斑海雀, PLATE CCCCXXX
Marsh wren, 长嘴沼泽鹪鹩, PLATE 98
Merlin, 灰背隼, PLATES 75, 92
Mississippi kite, 密西西比灰鸢, PLATE CXVII
Mountain bluebird, 山蓝鸲, PLATE CCCXCIII
Mountain plover, 岩鸻, PLATE CCCL
Mountain quail, 山翎鹑, PLATE CCCCXXIII
Mourning dove, 哀鸽, PLATE 17

Nashville warbler, 黄喉虫森莺, PLATE 89
Northern bobwhite (also red-shouldered hawk), 山齿鹑, PLATE 76
Northern cardinal, 主红雀, PLATE CLIX
Northern flicker, 北扑翅䴕, PLATES 37, CCCCXVI
Northern fulmar, 暴风鹱, PLATE CCLXIV
Northern gannet, 北鲣鸟, PLATE CCCXXVI
Northern goshawk, 苍鹰, PLATE CXLI
Northern hawk-owl, 猛鸮, PLATE CCCLXXVIII
Northern mockingbird, 小嘲鸫, PLATE 21
Northern parula, 北森莺, PLATE XV
Northern pygmy owl, 北美鸺鹠, PLATE CCCCXXXII
Northern pintail, 针尾鸭, PLATE CCXXVII
Northern saw-whet owl, 棕榈鬼鸮, PLATE CXCIX
Northern shoveler, 琵嘴鸭, PLATE CCCXXVII
Northern waterthrush, 黄眉灶莺, PLATE CCCCXXXIII
Olive-sided flycatcher, 绿胁绿霸鹟, PLATE CLXXIV, CCCCXXXIV
Orange-crowned warbler, 橙冠虫森莺, PLATE CLXXVIII
Orchard oriole, 圃拟鹂, PLATE 42
Osprey, 鹗, PLATE 81
Ovenbird, 橙顶灶莺, PLATE CXLIII

Painted bunting, 丽彩鹀, PLATE 53
Palm warbler, 棕榈林莺, PLATES CXLV, CLXIII
Passenger pigeon, 旅鸽（已灭绝）, PLATE 62
Pectoral sandpiper, 斑胸滨鹬, PLATE CCXCIV
Pelagic cormorant, 海鸬鹚, PLATE CCCCXII
Peregrine falcon, 游隼, PLATE 16
Pied-billed grebe, 斑嘴巨䴙䴘, PLATE CCXLVIII
Pileated woodpecker, 北美黑啄木鸟, PLATE CXI
Pine grosbeak, 松雀, PLATE CCCLVIII
Pine siskin, 松金翅雀, PLATE CLXXX
Pine warbler, 松莺, PLATE 30, CXL
Piping plover, 笛鸻, PLATE CCXX
Pomarine skua, 中贼鸥, PLATE CCLIII
Prairie warbler, 草原林莺, PLATE 14
Prothonotary warbler, 蓝翅黄森莺, PLATE III
Purple finch, 紫朱雀, PLATE IV
Purple martin, 紫崖燕, PLATE 22
Purple sandpiper, 紫滨鹬, PLATE CCLXXXIV
Pygmy nuthatch, 褐头䴓, PLATE CCCCXV

Razorbill, 刀嘴海雀, PLATE CCXIV
Red crossbill, 红交嘴雀, PLATE CXCVII
Red knot, 红腹滨鹬, PLATE CCCXV
Red-breasted nuthatch, 红胸䴓, PLATE CV
Red-cockaded woodpecker, 红顶啄木鸟, PLATE CCCLXXXIX
Reddish egret, 棕颈鹭, PLATE CCLVI
Red-bellied woodpecker, 红腹啄木鸟, PLATE CCCCXVI
Red-breasted merganser, 红胸秋沙鸭, PLATE CCCCI
Red-breasted sapsucker, 红胸吸汁啄木鸟, PLATE CCCCXVI
Red-eyed vireo, 红眼莺雀, PLATES CL, CCCCXXXIV

Redhead, 美洲潜鸭, PLATE CCCXXII
Red-headed woodpecker, 红头啄木鸟, PLATE 27
Red-necked grebe, 赤颈鸊鷉, PLATE CCXCVIII
Red-necked phalarope, 红颈瓣蹼鹬, PLATE CCXV
Red-shouldered hawk, 赤肩鵟, PLATES 56, 71
Red-tailed hawk, 红尾鵟, PLATES 51, 86
Red-throated diver, 红喉潜鸟, PLATE CCII
Red-winged blackbird, 红翅黑鹂, PLATE 67, CCCCXX
Rhinoceros auklet, 角嘴海雀, PLATE CCCCII
Ring-billed gull, 环嘴鸥, PLATE CCXII
Ring-necked duck, 环颈潜鸭, PLATE CCXXXIV
Rock ptarmigan, 岩雷鸟, PLATES CCCLXVIII, CCCCXVIII
Rock wren, 岩异鹪, PLATE CCCLX
Roseate spoonbill, 粉红琵鹭, PLATE CCCXXI
Roseate tern, 粉红燕鸥, PLATE CCXL
Rose-breasted grosbeak, 玫胸白斑翅雀, PLATE CXXVII
Rough-legged buzzard, 毛脚鵟, PLATES CLXVI, CCCCXXII
Royal tern, 橙嘴凤头燕鸥, PLATE CCLXXIII
Ruby-crowned kinglet, 红冠戴菊, PLATE CXCV
Ruby-throated hummingbird, 红喉北蜂鸟, PLATE 47
Ruddy duck, 棕硬尾鸭, PLATE CCCXLIII
Ruddy turnstone, 翻石鹬, PLATE CCCIV
Ruffed grouse, 披肩榛鸡, PLATE 41
Rufous hummingbird, 棕煌蜂鸟, PLATE CCCLXXIX
Rusty blackbird, 锈色黑鹂, PLATE CLVII
Sabine's gull and sanderling, 叉尾鸥, PLATE CCLXXXV
Sage thrasher, 高山弯嘴嘲鸫, PLATE CCCLXIX
Saltmarsh sharp-tailed sparrow, 尖尾沙鹀, PLATE CXLIX
Sanderling, 三趾滨鹬, PLATE CCXXX
Sandhill crane, 沙丘鹤, PLATE CCLXI
Sandwich tern, 白嘴端凤头燕鸥, PLATE CCLXXIX
Savannah sparrow, 稀树草鹀, PLATE CIX
Say's phoebe, 棕腹长尾霸鹟, PLATE CCCLIX
Scarlet ibis, 美洲红鹮, PLATE CCCXCVII
Scarlet tanager, 猩红丽唐纳雀, PLATE CCCLIV
Scissor-tailed flycatcher, 剪尾王霸鹟, PLATE CCCLIX
Seaside sparrow, 海滨沙鹀, PLATES 93, CCCLV
Sedge wren, 短嘴沼泽鹪鹩, PLATE CLXXV
Semipalmated plover, 半蹼鸻, PLATE CCCXXX
Semipalmated sandpiper, 半蹼滨鹬, PLATE CCCCV
Sharp-shinned hawk, 纹腹鹰, PLATE CCCLXXIV
Sharp-tailed grouse, 尖尾松鸡, PLATE CCCLXXXII
Short-billed dowitcher, 短嘴半蹼鹬, PLATE CCCXXXV
Short-eared owl, 短耳鸮, PLATE CCCCXXXII
Small-headed flycatcher, 小头莺, PLATE CCCCXXXIV
Smew, 斑头秋沙鸭, PLATE CCCXLVII
Smith's longspur, 黄腹铁爪鹀, PLATE CCCC
Snow bunting, 雪鹀, PLATE CLXXXIX
Snow goose, 雪雁, PLATE CCCLXXXI
Snowy egret, 雪鹭, PLATE CCXLII
Snowy owl, 雪鸮, PLATE CXXI
Solitary sandpiper, 褐腰草鹬, PLATE CCLXXXIX
Song sparrow, 歌带鹀, PLATES 25, CCCXC
Sooty albatross, 乌信天翁, PLATE CCCCVII
Sooty tern, 乌燕鸥, PLATE CCXXXV
Sora rail, 黑脸田鸡, PLATE CCXXXIII
Spotted sandpiper, 斑腹矶鹬, PLATE CCCX
Spotted towhee, 棕胁唧鹀, PLATE CCCXCIV
Spruce grouse, 枞树镰翅鸡, PLATE CLXXVI
Steller's eider, 小绒鸭, PLATE CCCCXXIX
Steller's jay, 暗冠蓝鸦, PLATE CCCLXII
Stilt-sandpiper, 高跷鹬, PLATE CCCXLIV
Summer tanager, 玫红丽唐纳雀, PLATE 44
Surfbird, 短嘴鹬, PLATE CCCCXXVIII
Surf scoter, 斑头海番鸭, PLATE CCCXVII
Swainson's hawk, 斯氏鵟, PLATE CCCLXXII
Swainson's warbler, 白眉食虫莺, PLATE CXCVIII
Swallow-tailed kite, 燕尾鸢, PLATE 72
Swamp sparrow, 沼泽带鹀, PLATE 64

Tengmalm's owl, 鬼鸮, PLATE CCCLXXX
Tennessee warbler, 灰冠虫森莺, PLATE CLIV
Three-toed woodpecker, 三趾啄木鸟（美洲）, PLATE CCCCXVII
Townsend's bunting, 唐氏雀, PLATE CCCC
Townsend's solitaire, 坦氏孤鸫, PLATE CCCCXIX
Townsend's warbler, 黄眉林莺, PLATE CCCXCIII
Tree swallow, 双色树燕, PLATE 100
Tricolored heron, 三色鹭, PLATE CCXVII
Tricoloured blackbird, 三色黑鹂, PLATE CCCLXXXVIII
Trudeau's tern, 白顶燕鸥, PLATE CCCCIX
Trumpeter swan, 黑嘴天鹅, PLATES CCCLXXVI, CCCCVI
Tufted puffin, 簇羽海鹦, PLATE CCXLIX
Tufted titmouse, 美洲凤头山雀, PLATE 39
Tundra swan, 小天鹅, PLATE CCCCXI
Turkey-vulture, 红头美洲鹫, PLATE CLI
Upland sandpiper, 高原鹬, PLATE CCCIII

Varied thrush, 杂色鸫, PLATES CCCLXIX, CCCCXXXIII
Veery, 棕夜鸫, PLATE CLXIV
Velvet scoter, 斑脸海番鸭, PLATE CCXLVII
Vesper sparrow, 栗肩雀鹀, PLATE 94
Violet-green swallow, 紫绿树燕, PLATE CCCLXXXV
Virginia rail, 弗吉尼亚秧鸡, PLATE CCV

Warbling vireo, 歌莺雀, PLATE CXVIII
Water pipit, 黄腹鹨, PLATES, X, 80
Western bluebird, 西蓝鸲, PLATE CCCXCIII
Western kingbird, 西王霸鹟, PLATE CCCLIX
Western scrub jay, 西丛鸦, PLATE CCCLXII
Western tanager, 黄腹丽唐纳雀, PLATES CCCLIV, CCCC
Whimbrel, 中杓鹬, PLATE CCXXXVII
Whip-poor-will, 三声夜鹰, PLATE 82
White ibis, 美洲白鹮, PLATE CCXXII
White-breasted nuthatch, 白胸鳾, PLATE CLII
White-crowned pigeon, 白顶鸽, PLATE CLXXVII
White-crowned sparrow, 白冠带鹀, PLATE CXIV
White-eyed vireo, 白眼莺雀, PLATE 63
White-rumped sandpiper, 白腰滨鹬, PLATE CCLXXVIII
White-tailed kite, 黑翅鸢, PLATE CCCLII
White-tailed ptarmigan, 白尾雷鸟, PLATE CCCCXVIII
White-tailed tropicbird, 白尾鹲, PLATE CCLXII
White-throated sparrow, 白喉带鹀, PLATE VIII
White-winged crossbill, 白翅交嘴雀, PLATE CCCLXIV
Whooping crane, 美洲鹤, PLATE CCXXVI
Wild turkey, 火鸡, PLATES I, VI
Willet, 斑翅鹬, PLATE CCLXXIV
Willow flycatcher, 桤木纹霸鹟, PLATE 45
Willow grouse, 柳雷鸟, PLATE CXCI
Wilson's phalarope, 细嘴瓣蹼鹬, PLATE CCLIV

Wilson's plover, 厚嘴鸻, PLATE CCIX
Wilson's storm petrel, 黄蹼洋海燕, PLATE CCLXX
Wilson's warbler, 黑头威森莺, PLATE CXXIV
Winter wren, 鹪鹩, PLATE CCCLX
Wood duck, 林鸳鸯, PLATE CCVI
Wood stork, 黑头鹮鹳, PLATE CCXVI
Wood thrush, 棕林鸫, PLATE 73
Worm-eating warbler, 食虫莺, PLATE 34
Yellow rail, 北美花田鸡, PLATE CCCXXIX
Yellow warbler, 黄林莺, PLATES 35, 65, 95
Yellow-bellied sapsucker, 黄腹吸汁啄木鸟, PLATE CXC
Yellow-billed cuckoo, 黄嘴美洲鹃, PLATE II
Yellow-billed magpie, 黄嘴喜鹊, PLATE CCCLXII
Yellow-breasted chat, 黄胸大鹏莺, PLATE CXXXVII
Yellow-crowned night heron, 黄冠夜鹭, PLATE CCCXXXVI
Yellow-headed blackbird, 黄头黑鹂, PLATE CCCLXXXVIII
Yellow-rumped warbler, 黄腰林莺, PLATES CLIII, CCCXCV
Yellow-throated vireo, 黄喉莺雀, PLATE CXIX, CCCCXXXIV
Yellow-throated warbler, 黄喉林莺, PLATE 85

Zenaida dove, 鸣哀鸽, PLATE CLXII